Design of Optical Interference Coatings

McGraw-Hill Optical and Electro-Optical Engineering Series

Robert E. Fischer and Warren J. Smith, *Series Editors*

Published

Nishihara, Haruna, Suhara • *Optical Integrated Circuits*
Rancourt • *Optical Thin Films Users' Handbook*

Forthcoming Volumes

Fantone • *Optical Fabrication*
Fischer • *Optical Design*
Gilmore • *Expert Vision Systems*
Jacobson • *Thin Film Deposition*
Johnson • *Infrared System Design*
Ross • *Laser Communications*
Stover • *Optical Scattering*
Vukobratovich • *Introduction to Opto-Mechanical Design*

Other Books of Interest

CSELT • *Optical Fibre Communication*
Hecht • *The Laser Guidebook*
Hewlett-Packard • *Optoelectronics Applications Manual*
Kao • *Optical Fiber Systems*
Macleod • *Thin Film Optical Filters*   0-44674
Marshall • *Free Electron Lasers*
Optical Society of America • *Handbook of Optics*
Smith • *Modern Optical Engineering*
Texas Instruments • *Optoelectronics*
Wyatt • *Radiometric System Design*

For more information about other McGraw-Hill materials, call 1-800-2-MCGRAW in the United States. In other countries, call your nearest McGraw-Hill office.

624729

1-800-262-4729

Design of Optical Interference Coatings

ALFRED THELEN

Vice President
Thin Film Research
Leybold AG
Hanau, West Germany

McGraw-Hill Book Company

New York   St. Louis   San Francisco   Auckland
Bogotá   Hamburg   London   Madrid   Milan   Mexico
Montreal   New Delhi   Panama   Paris   São Paulo
Singapore   Sydney   Tokyo   Toronto

Library of Congress Cataloging-in-Publication Data

Thelen, Alfred.
  Design of optical interference coatings/Alfred Thelen.
    p. cm.
  Bibliography: p.
  Includes index.
  ISBN 0-07-063786-5
  1. Optical coatings. I. Title.
TS517.2.T47 1988b
681'.4—dc19   88-19152
                                                  CIP

1234567890   DOCDOC   895432109

ISBN 0-07-063786-5

The editors for this book were Daniel A. Gonneau and Susan Thomas, the designer was Naomi Auerbach, and the production supervisor was Suzanne W. Babeuf. It was set in Century Schoolbook by J. M. Post Graphics, Corp.

Printed and bound by the R. R. Donnelley and Sons Company.

For more information about other McGraw-Hill materials, call 1-800-2-MCGRAW in the United States. In other countries, call your nearest McGraw-Hill office.

Contents

•

Preface

This in-depth treatise on the design of optical interference coatings is intended to be a *source of new ideas* for the practicing optical coating designer, a *textbook* for the advanced students of optical engineering, and a *reference* for those who recognize the key role optical coatings play in optical instruments, lasers and laser systems, fiber communication links, optical recording/storage heads and media, display systems, infrared guidance and detection devices, photoelectric converters, architectural glass, eyeglasses, etc., and who would like to understand the trade-offs which go into their design.

After a brief introduction into the *tasks* and *philosophy* of design in chapter 1, we derive, in the mathematically heavier chapters 2 and 3, the *basic relationships* and *design methods*. Such topics as characteristic matrices, periodic structures, equivalent layers, Chebyshev prototypes, effective interfaces, and buffer layers are covered.

In chapters 4 through 10, specific coating types are discussed: antireflection coatings, high reflectors, neutral beam splitters, edge filters, minus filters, polarizers, wide and narrow bandpasses, and nonpolarizing coatings.

In chapter 11 on computer refining, we go into applying the *final polish* to the designs developed in the earlier chapters. How *deviations* from the theoretical assumptions affect the practical performance is the topic of chapter 12 on producibility of designs.

Many designs, models, and derivations appear in this book *for the first time*. They are partly the result of the author's being forced to think through all aspects of design and partly an opportunity to present designs which rested in the author's files waiting to be published.

One might wonder whether, in a book of this kind, it is really necessary to derive most formulas. The answer is yes. In order to judge the applicability and the limits of a design model, one has to know the assumptions and limits of the theory behind it. If the derivations were not presented here, the reader would have to look them up in the literature, where each publication uses a different nomenclature and a different style of derivation (geometric vs. matrices vs. recursion formulas, and so on).

Literature references are made on the basis of either who published first or who made a presentation best suited to the style of the book.

Various computer programs were used to calculate the data in support of the theory. The programs were written in BASIC for use on Hewlett-Packard Model 9000 Series 200 computers. Some of the programs are listed under "Problems and Solutions." In order to facilitate debugging, the input data of a sample calculation are incorporated in "DATA" statements.

Most figures were generated on a Hewlett-Packard Model 7470A Plotter.

The author, an intuitive thinking type, hopes that the reader will forgive him that the search for errors had to stop at some reasonable point of diminishing returns.

Alfred Thelen

Acknowledgments

With gratitude I dedicate this book to two outstanding industrial pioneers

Dr. -Ing. Alfred Hauff
President, Leybold AG
Hanau, West Germany

Rolf Illsley
Chairman of the Board
Optical Coating Laboratory, Inc.
Santa Rosa, California

Without their encouragement and generous support, this book would not have come into existence.

The material presented in this book was developed during 30 years of design practice. Many colleagues helped in the crystallization process, especially Joseph Apfel, Philip Baumeister, Peter Berning, Oliver Heavens, Carol Jacobs (Snavely), Angus Macleod, and Anthony Musset. Special thanks go to Joseph Apfel and Angus Macleod for helping with the tedious task of proofreading the manuscript.

Alfred Thelen
Hanau, West Germany

List of Symbols

A, B, M, Q, I, S	Two by two matrices, real elements in the principal diagonal, imaginary elsewhere
A, B, C, D, M, L, H	Quarter-wave-thick layers with refractive indices n_A, n_B, n_C, n_D, n_M, n_L, n_H
A_{11}, iA_{12}, iA_{21}, A_{22}, . . .	Matrix elements
$\overline{AB}$, $\overline{BC}$, $\overline{AD}$	Line between the points A and B, B and C, and A and D (Fig. 2.5)
A, A_1, B, C	Constants in Collin's formulas (Table 3.4)
C, C_1	Constants (Eqs. 3.33 and 3.35)
D	Denominator (Eq. 3.41)
$\vec{E}$	Electric field vector
$E(z)$	Magnitude of total electric field
$E^+(z)$, $E^-(z)$	Magnitude of resultant of all positive- and negative-going waves
ERAR	Equal ripple antireflection coating, Sec. 3.2.1, Fig. 3.10
ERF	Equal ripple filter, Sec. 3.2.2, Fig. 3.10
$\vec{H}$	Magnetic field vector
$H(z)$	Magnitude of total magnetic field
$H^+(z)$, $H^-(z)$	Magnitude of resultant of all positive- and negative-going waves
I	Unity matrix
$\vec{K}$	Propagation vector of electromagnetic wave
L	Eigenvalues (Eq. 2.29)
L	Last matrix (Eq. 3.33)
M	Characteristic matrix (Eq. 2.17)
M_{11}, iM_{12}, iM_{21}, M_{22}	Elements of characteristic matrix
N	Equivalent index (Eq. 3.1)
P	Polynomial (Eq. 3.22)
Q	Transfer matrix (Eq. 2.34)
Q_{11}, iQ_{12}, iQ_{21}, Q_{22}	Elements of transfer matrix
QWOT	Quarter-wave optical thickness (Eq. 2.56)
$\vec{R}$	Reflectance vector
R	$= \vec{R}\,\vec{R}^*$, reflectance
$S_m(x)$	Chebyshev polynomial of the first kind, mathematical version (Table 2.1)
$\vec{T}$	Transmittance vector

T	$= TT^*$, transmittance
$T_m(x)$	Chebyshev polynomial of the first kind, engineering version (Table 3.3)
W	Fractional bandwidth (Sec. 3.2.1)
X	Ratio of even and odd refractive indices
Z_0	$= \sqrt{\mu_0/\varepsilon_0}$, impedance of free space
$a, b, c, \ldots$	Fractional thickness (Eq. 2.57) coefficients of series: $a_0, a_1, \ldots$ (Eq. 3.21)
c	Constant (Eq. 2.44)
d	Physical thickness
i, j	Running indices
k	Number of points to be refined
	Order of stop band
m	Number of layers
n	Refractive index
n_0	Refractive index of incident medium
n_s	Refractive index of substrate
Δn	$= n_p/n_S$
p	As subscript, parallel plane of polarization
p	Number of periods (Eqs. 2.30 and 2.62)
$\overrightarrow{r}$	Reflectance vector in reverse direction
r	$= \sqrt{r_A R_B}$, term in split filter formula (Eq. 2.55)
s	As subscript, vertical plane of polarization
s	$= i \sin \phi / \cos \phi$ (Eq. 3.39)
u, v, w	Terms of equations
x	$= M_{11} + M_{22}$
x, y	Index ratios
z	Ordinate in the direction perpendicular to the film plane
Γ	Equivalent thickness (Eq. 3.2)
ϕ	Phase of the reflectance vector (Eq. 2.55)
Λ	Lagrangian multiplier (Eq. 11.6)
α	Angle between incident light and normal to film plane (α_0 in incident medium, α_m inside layer m)
β	Matching angle (Eq. 2.58)
δ	Small quatlity $\delta \rightarrow 0$
ϕ	Phase thickness (Eq. 2.5)
η	$= 1/\sin(\pi W/4)$ (Sec. 3.2.1)
λ	Wavelength
λ_0	Design wavelength
λ_0/λ	Relative wavenumber
θ	Phase of transmittance vector
ρ	Passband ripple of Chebyshev prototypes (Eq. 3.23)

Introduction

*There is nothing more practical than a good
theory.*   KURT LEWIN

*The purpose of computing is insight, not
numbers.*   R. W. HAMMING

*"What makes the desert beautiful," said the
little prince, "is that somewhere it hides a
well...."*   ANTOINE DE SAINT EXUPÉRY

The purpose of this book is teaching how to combine a sequence of
mainly nonabsorbing thin films with thicknesses comparable to the
wavelength of light with an optical interference coating which exhibits
one of the spectral characteristics shown in Fig. 1.1 (so-called *classical
optical coatings*). One may ask whether, with the power of today's
computers, the design of an optical coating is just a matter of putting
a specification into a computer and letting high speed and large mem-
ory do the rest: The bigger the computer the better the design?

It is the author's experience that it is seldom possible to translate a
design into a practical coating when all that is known is a set of thicknesses and indices of refraction. What makes a design work should be
in the head of the coating engineer and not in the "bowels" of the
computer (as Baumeister[1] puts it). The production of an optical coating
often requires a precision in the deposition of the individual layers
which surpasses the monitoring accuracy of the coating machine. For

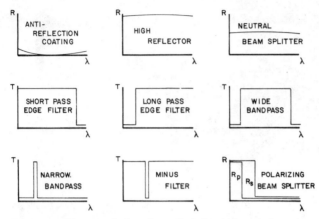

Figure 1.1. Characteristics of so-called classical optical coatings.

different designs, the final coating is the expansion of a simpler coating, based on the same model, tested in parts, and scaled up with increasing complexity. Often, the theoretical assumptions for the computations are not well realized: Just compare the cross section of a multilayer, as shown in Fig. 1.2, with the usual assumptions of plan parallelism, homogeneity, and isotropy. Also, if the computer fails to zero in on the desired characteristic (zero reflectance over a very wide wavelength region, for example) how can one decide whether the coating is impossible or the solution is just another 100 computer hours away.

The design of an optical coating, as taught in this book, is divided into the following steps:

Figure 1.2. Cross section of part of a ZnS-MgF$_2$ laser mirror. (*Guenther and Pulker.*[2])

1. Selection of a design method
2. Derivation of the core design
3. Improvement of the core design (smoothing the passband region, increasing the steepness of transition, eliminating holes in the stop band, etc.)
4. Adaptation to deviations from the theoretical model (dispersion, inhomogeneities, small absorption, scattering, etc.)
5. Computer refinement

The advantage of this design philosophy is that the computer is used as a "calculator" only. The designer and not the computer is in charge. With the application of optical models an inherent stability and easy scaleability is generated.

This book deals extensively with steps 1 to 3. Many general solutions are derived. For steps 4 and 5 only methods, tools, and sample solutions can be given. Each final solution depends on the specific application and method of manufacturing. Unless the opposite is specifically stated, the coatings are assumed to be free of absorption and dispersion. In Chaps. 4 and 5 some designs use metal films. The influence of dispersion, inhomogeneity, weak absorption, and thickness variations is discussed in Chap. 12.

Chapter

2

Theory

The treatment of multilayer film theory used here will deal with the resultant electric and magnetic fields and their boundary conditions in the various regions (Hecht and Zajac[1]). We will be dealing with two types of wave functions, those differentiating between positive- (we take as positive the direction of the incident light) and negative-going waves [$\vec{E}^+(z)$ and $\vec{E}^-(z)$] and those which do not [$\vec{E}(z)$ and $\vec{H}(z)$]. For example, $\vec{E}^+(z_1)$ represents the resultant of all waves at $z = z_1$ traveling in the positive direction and $\vec{E}(z_2)$ represents the resultant of all waves at $z = z_2$ traveling in both directions. Equations between the first type of wave functions in the massive media on both sides of the multilayer will lead to the transfer matrix and formulas for the reflectance and transmittance of the multilayer. Equations between the second type of wave functions will lead to the characteristic matrix of a multilayer which only depends on the optical parameters of the multilayer and not on the optical characteristics of the massive media surrounding it (Berning[2]).

Consider the linearly polarized wave shown in Fig. 2.1 impinging on a thin film (assumed to be nonabsorbing, homogeneous, isotropic, and of uniform thickness) with refractive index n between two massive transparent media with refractive indices n_0 and n_s. $\vec{E}$ is assumed to be perpendicular to the incidence plane (s or perpendicular polarization).

The boundary condition to the laws of electromagnetic wave theory (Maxwell's equations) requires that the tangential components of both the electric ($\vec{E}$) and magnetic ($\vec{H}$) field vectors be continuous across the boundaries (i.e., equal on both sides, the two sides are described by $z - \delta$ and $z + \delta$ with $\delta \to 0$). At boundary $z = z_1$

$$E(z_1) = E^+(z_1 - \delta) + E^-(z_1 - \delta) = E^+(z_1 + \delta) + E^-(z_1 + \delta)$$
(2.1)

Also,

$$Z_0 H(z_1) = [E^+(z_1 - \delta) - E^-(z_1 - \delta)]n_0 \cos \alpha_0 \qquad (2.2)$$
$$= [E^+(z_1 + \delta) - E^-(z_1 + \delta)]n \cos \alpha$$

where $Z_0 = \sqrt{\mu_0/\varepsilon_0}$ is the impedance of free space
   (377 Ω in mks units)

$\alpha_0 =$ incidence angle (defined as the angle between the propagation vector $\vec{K}$ and the normal to the film surface)

$\alpha =$ angle of wave propagation in the film as determined from Snell's law ($n_0 \sin \alpha_0 = n \sin \alpha$)

Use is made of the fact that the vectors $\vec{E}$ and $\vec{H}$ in nonmagnetic media are related by the refractive index and the unit propagation vector $\vec{K}$

$$Z_0 \vec{H} = n\vec{K} \times \vec{E}$$

Since there is no reflected wave in the exit medium we have at $z = z_2$

$$E(z_2) = E^+(z_2 - \delta) + E^-(z_2 - \delta) = E^+(z_2 + \delta) \qquad (2.3)$$

and

$$Z_0 H(z_2) = [E^+(z_2 - \delta) - E^-(z_2 - \delta)]n \cos \alpha \qquad (2.4)$$
$$= E^+(z_2 + \delta)n_s \cos \alpha_s$$

A wave which traverses the film once undergoes a shift in phase of

$$\phi = \frac{2\pi}{\lambda} nd \cos \alpha \qquad (2.5)$$

E ⊥ INCIDENCE PLANE

OR S-POLARIZATION (T$\mathcal{E}$)

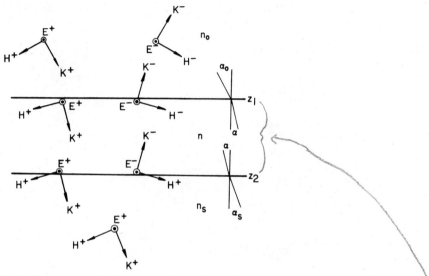

Figure 2.1. Assumed electromagnetic fields at the boundaries of a thin film. $\vec{E}$ is perpendicular to the plane of incidence.

where $d = z_2 - z_1$ is the physical thickness of the film and ϕ is called the angular phase thickness. (A geometrical explanation of Eq. 2.5 is given in Fig. 2.5 of Sec. 2.10.) Equation 2.5 allows us to establish the following relationships:

$$E^+(z_2 - \delta) = E^+(z_1 + \delta)e^{-i\phi} \quad \text{and} \quad E^-(z_2 - \delta) = E^-(z_1 + \delta)e^{+i\phi}$$

with $i = \sqrt{-1}$. Equations 2.3 and 2.4 can now be written as

$$E(z_2) = E^+(z_1 + \delta)e^{-i\phi} + E^-(z_1 + \delta)e^{+i\phi}$$

and

$$Z_0 H(z_2) = [E^+(z_1 + \delta)e^{-i\phi} - E^-(z_1 + \delta)e^{+i\phi}]n \cos \alpha$$

The last two equations can be solved for $E^+(z_1 + \delta)$ and $E^-(z_1 + \delta)$:

$$E(z_2) + \frac{Z_0 H(z_2)}{n \cos \alpha} = 2E^+(z_1 + \delta)e^{-i\phi}$$

and

and

$$E(z_2) - \frac{Z_0 H(z_2)}{n \cos \alpha} = -2E^-(z_1 + \delta)e^{i\phi}$$

which, when substituted into Eqs. 2.1 and 2.2, yield

$$E(z_1) = E(z_2) \cos \phi + \frac{Z_0 H(z_2) (i \sin \phi)}{n \cos \alpha} \tag{2.6}$$

and

$$Z_0 H(z_1) = E(z_2) i(n \cos \alpha) \sin \phi + Z_0 H(z_2) \cos \phi \tag{2.7}$$

using $\cos \phi = (e^{i\phi} + e^{-i\phi})/2$ and $\sin \phi = (e^{i\phi} - e^{-i\phi})/2$.

Let us now assume $\vec{E}$ to be parallel to the plane of incidence (p or parallel polarization, Fig. 2.2). Then, Eqs. 2.1 and 2.2 become

$$\begin{aligned} E(z_1) &= [E^+(z_1 - \delta) + E^-(z_1 - \delta)] \cos \alpha_0 \\ &= [E^+(z_1 + \delta) + E^-(z_1 + \delta)] \cos \alpha \end{aligned} \tag{2.8}$$

and

$$\begin{aligned} Z_0 H(z_1) &= n_0 [E^+(z_1 - \delta) - E^-(z_1 - \delta)] \\ &= n[E^+(z_1 + \delta) - E^-(z_1 + \delta)] \end{aligned} \tag{2.9}$$

Equations 2.3 and 2.4 become

$$\begin{aligned} E(z_2) &= [E^+(z_2 - \delta) + E^-(z_2 - \delta)] \cos \alpha_0 \\ &= [E^+(z_1 + \delta)e^{-i\phi} + E^-(z_1 + \delta)e^{i\phi}] \cos \alpha \end{aligned} \tag{2.10}$$

and

$$\begin{aligned} Z_0 H(z_2) &= n[E^+(z_2 - \delta) - E^-(z_2 - \delta)] \\ &= n[E^+(z_1 + \delta)e^{-i\phi} - E^-(z_1 + \delta)e^{i\phi}] \end{aligned} \tag{2.11}$$

Adding and subtracting the last two equations yields

E ‖ INCIDENCE PLANE

OR P-POLARIZATION (TM)

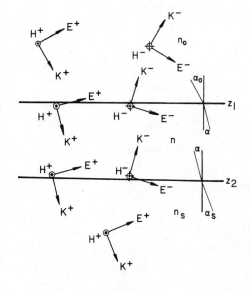

Figure 2.2. Assumed electromagnetic fields at the boundaries of a thin film. $\vec{E}$ is parallel to the plane of incidence.

$$\frac{E(z_2)}{\cos \alpha} + \frac{Z_0 H(z_2)}{n} = 2E^+(z_1 + \delta)e^{-i\phi}$$

and

$$\frac{E(z_2)}{\cos \alpha} - \frac{Z_0 H(z_2)}{n} = 2E^-(z_1 + \delta)e^{i\phi}$$

We now can eliminate $E^+(z_1 + \delta)$ and $E^-(z_1 + \delta)$ in Eqs. 2.8 and 2.9 and obtain similarly as before

$$E(z_1) = E(z_2) \cos \phi + \frac{Z_0 H(z_2)(i \sin \phi)}{n/\cos \alpha} \tag{2.12}$$

and

$$Z_0 H(z_1 = E(z_2) \, i\left(\frac{n}{\cos \alpha}\right) \sin \phi + Z_0 H(z_2)\cos \phi \tag{2.13}$$

In-between orientations of $\vec{E}$ are determined by vector addition of the perpendicular and parallel components.

We note that Eqs. 2.6 and 2.7 can be transformed into Eqs. 2.12 and

2.13 by replacing $n \cos \alpha$ with $n/\cos \alpha$. We can consequently build our theory on two equations rather than four:

$$E(z_1) = E(z_2) \cos \phi + \frac{Z_0 H(z_2)(i \sin \phi)}{n} \qquad (2.14)$$

and

$$Z_0 H(z_1) = E(z_2) in \sin \phi + Z_0 H(z_2) \cos \phi \qquad (2.15)$$

by adhering to the following convention:

$$n \Rightarrow n_s = n \cos \alpha \qquad \text{for } s \text{ or perpendicular polarization} \quad TL$$

$$n \Rightarrow n_p = \frac{n}{\cos \alpha} \qquad \text{for } p \text{ or parallel polarization} \quad TM \quad (2.16)$$

$$n \Rightarrow n \qquad \text{for normal light incidence}$$

See McLeod P.96 §§.21-36

The price for this simplification is the ambiguity in the use of the symbol n. n now can mean the refractive index as measured with a refractometer or the "effective index" as defined in Eqs. 2.16 (Knittl[3]).

2.1. Characteristic Matrix of a Single Film

In matrix notation, Eqs. 2.14 and 2.15 can be written in the following form:

$$\begin{bmatrix} E(z_1) \\ Z_0 H(z_1) \end{bmatrix} = \begin{bmatrix} \cos \phi & \frac{i \sin \phi}{n} \\ in \sin \phi & \cos \phi \end{bmatrix} \begin{bmatrix} E(z_2) \\ Z_0 H(z_2) \end{bmatrix}$$

The matrix

$$\mathbf{M} = \begin{bmatrix} \cos \phi & \frac{i \sin \phi}{n} \\ in \sin \phi & \cos \phi \end{bmatrix} \qquad (2.17)$$

is called the characteristic matrix of the film. It contains only parameters of the film.

Two special cases of the single film are of considerable practical importance:

1. *The quarter-wave layer.*   When $nd \cos \alpha = \lambda_0/4$ then $\phi = \pi/2$ and

$$\mathbf{M}_{\lambda/4} = \begin{bmatrix} 0 & \dfrac{i}{n} \\ in & 0 \end{bmatrix} \tag{2.18}$$

2. *The half-wave layer.*   When $nd \cos \alpha = \lambda_0/2$ then $\phi = \pi$ and

$$\mathbf{M}_{\lambda/2} = \begin{bmatrix} -1 & 0 \\ 0 & -1 \end{bmatrix} \tag{2.19}$$

2.2. Characteristic Matrix of a Sequence of Layers

The characteristic matrix (Eq. 2.17) relates the fields at the boundary $z = z_1$ to the fields at boundary $z = z_2$. It follows, therefore, that if two overlaying films were deposited on the substrate, there would be three boundaries ($z = z_1$, $z = z_2$, and $z = z_3$), and now

$$\begin{bmatrix} E(z_2) \\ Z_0H(z_2) \end{bmatrix} = \mathbf{M}_2 \begin{bmatrix} E(z_3) \\ Z_0H(z_3) \end{bmatrix}$$

Premultiplying both sides with $\mathbf{M}_1$, we obtain

$$\begin{bmatrix} E(z_1) \\ Z_0H(z_1) \end{bmatrix} = \mathbf{M}_1\mathbf{M}_2 \begin{bmatrix} E(z_3) \\ ZH(z_3) \end{bmatrix}$$

In general, if m is the number of layers then the first ($z = z_1$) and last ($z = z_{m+1}$) boundary are related by

$$\begin{bmatrix} E(z_1) \\ Z_0H(z_1) \end{bmatrix} = \mathbf{M}_1\mathbf{M}_2 \cdots \mathbf{M}_m \begin{bmatrix} E(z_{m+1}) \\ Z_0H(z_{m+1}) \end{bmatrix} \tag{2.20}$$

We see that the characteristic matrix of a sequence of layers is the resultant of the product of the individual characteristic matrices

$$\mathbf{M} = \mathbf{M}_1 \, \mathbf{M}_2 \, \mathbf{M}_3 \cdots \mathbf{M}_m \tag{2.21}$$

The characteristic matrix of a single layer (Eq. 2.17) has real elements in the principal diagonal and purely imaginary elements elsewhere. Let us now multiply two matrices $\mathbf{A}$ and $\mathbf{B}$ of this type

$$\begin{bmatrix} A_{11} & iA_{12} \\ iA_{21} & A_{22} \end{bmatrix} \begin{bmatrix} B_{11} & iB_{12} \\ iB_{21} & B_{22} \end{bmatrix}$$

$$= \begin{bmatrix} A_{11}B_{11} - A_{12}B_{21} & i(A_{11}B_{12} + A_{12}B_{22}) \\ i(A_{21}B_{11} - A_{22}B_{21}) & -A_{21}B_{12} + A_{22}B_{22} \end{bmatrix} = \begin{bmatrix} M_{11} & iM_{12} \\ iM_{21} & M_{22} \end{bmatrix} \tag{2.22}$$

We see that the product matrix **M** is of the same type as **A** and **B**. We can conclude that the characteristic matrix of a sequence of layers also has real elements on the principal diagonal and purely imaginary elements elsewhere.

The most significant feature of this result is the fact that one can represent any sequence of layers by a 2 by 2 matrix completely independent of the parameters of the surrounding media.

2.3. Determinant of the Characteristic Matrix

The determinant of the characteristic matrix of a single layer (Eq. 2.17) is

$$\det(\mathbf{M}) = \cos\phi\cos\phi + \sin\phi\sin\phi = 1$$

We ask now what is the value of the characteristic matrix of a sequence of layers? Since the multiplication of matrices is performed the same way as the multiplication of determinants, it follows that for square matrices **A** and **B** with an equal number of rows (Kreyszig[4])

$$\det(\mathbf{AB}) = \det(\mathbf{A})\det(\mathbf{B}) \cdot$$

If we apply this result to Eq. 2.21 we find

$$\det(\mathbf{M}) = \det(\mathbf{M}_1)\det(\mathbf{M}_2)\det(\mathbf{M}_3)\cdots\det(\mathbf{M}_m) \qquad (2.23)$$
$$= \mathbf{M}_{11}\mathbf{M}_{22} + \mathbf{M}_{12}\mathbf{M}_{21} = 1$$

or in words: The *determinant* of the characteristic matrix of a sequence of layers always equals 1.

2.4. Stack Reversal

It is not necessary to recompute the characteristic matrix of a sequence of layers when the order of the layers is reversed. Due to the fact that the inverse of a product of matrices equals the product of the inverse individual matrices in reverse order (Kreyszig[4]), we can conclude from Eq. 2.21 that

$$\mathbf{M}^{-1} = \mathbf{M}_m{}^{-1}\,\mathbf{M}_{m-1}{}^{-1}\cdots\mathbf{M}_2{}^{-1}\,\mathbf{M}_1{}^{-1} \qquad (2.24)$$

The inverse of the characteristic matrix of a sequence of layers is

$$\mathbf{M}^{-1} = \begin{bmatrix} M_{11} & iM_{12} \\ iM_{21} & M_{22} \end{bmatrix}^{-1} = \begin{bmatrix} M_{22} & -iM_{12} \\ -iM_{21} & M_{11} \end{bmatrix} \qquad (2.25)$$

and of a single film (Eq. 2.17)

$$\begin{bmatrix} \cos\phi & \dfrac{i\sin\phi}{n} \\ in\sin\phi & \cos\phi \end{bmatrix}^{-1} = \begin{bmatrix} \cos\phi & \dfrac{-i\sin\phi}{n} \\ -in\sin\phi & \cos\phi \end{bmatrix}$$

$$M^{-1} = M^{*}$$

The right side of this equation can also be generated by pre- and post-multiplying the characteristic matrix (Eq. 2.17) of the single film with the matrix (Thelen[5])

$$\mathbf{S} = \begin{bmatrix} 1 & 0 \\ 0 & -1 \end{bmatrix}$$

since

$$\begin{bmatrix} \cos\phi & \dfrac{-i\sin\phi}{n} \\ -in\sin\phi & \cos\phi \end{bmatrix}$$

$$= \begin{bmatrix} 1 & 0 \\ 0 & -1 \end{bmatrix} \begin{bmatrix} \cos\phi & \dfrac{i\sin\phi}{n} \\ in\sin\phi & \cos\phi \end{bmatrix} \begin{bmatrix} 1 & 0 \\ 0 & -1 \end{bmatrix} \quad (2.26)$$

Using Eq. 2.26 for all right-side matrices and noting that $\mathbf{SS} = \mathbf{I}$, Eq. 2.24 becomes

$$\mathbf{M}^{-1} = \mathbf{S}\mathbf{M}_m\mathbf{M}_{m-1} \cdots \mathbf{M}_2\mathbf{M}_1\mathbf{S}$$

and, after pre- and postmultiplying both sides with $\mathbf{S}$,

$$\mathbf{SM}^{-1}\mathbf{S} = \begin{bmatrix} M_{22} & iM_{12} \\ iM_{21} & M_{11} \end{bmatrix} = \mathbf{M}_m\mathbf{M}_{m-1} \cdots \mathbf{M}_2\mathbf{M}_1 = \mathbf{M}_{\mathrm{rev}} \quad (2.27)$$

In words: Given the characteristic matrix of a sequence of layers, the characteristic matrix of the same multilayer arranged in reverse order can be obtained by *exchanging* the elements of the *principal diagonal*.

2.5. Symmetric Layer Sequences

A symmetric layer sequence can be imagined consisting of two parts: a certain sequence of layers with matrix $\mathbf{A}$ and the same sequence arranged in reverse order (matrix $\mathbf{A}_{\mathrm{rev}}$). Using Eq. 2.27 we then can write for the characteristic matrix of the symmetric layer sequence $\mathbf{M}_{\mathrm{sym}}$ in terms of the matrix elements of the half-sequence A_{11}, A_{12}, A_{21}, and A_{22}:

$$\mathbf{M}_{\mathrm{sym}} = \begin{bmatrix} A_{11} & iA_{12} \\ iA_{21} & A_{22} \end{bmatrix} \begin{bmatrix} A_{22} & iA_{12} \\ iA_{21} & A_{11} \end{bmatrix}$$

$$= \begin{bmatrix} A_{11}A_{22} - A_{12}A_{21} & 2iA_{11}A_{12} \\ 2iA_{21}A_{22} & A_{11}A_{22} - A_{12}A_{21} \end{bmatrix} \quad (2.28)$$

We consequently find that a symmetric layer sequence has equal elements on the principal diagonal.

2.6. Periodic Multilayers

In a periodic multilayer the layers making up a base period are repeated p times. If the characteristic matrix of the base is $\mathbf{M}$, Eq. 2.15 demands that the characteristic matrix of the periodic multilayer is $\mathbf{M}^p$. With the use of Chebyshev polynomials, $\mathbf{M}^p$ can be calculated very efficiently (Abelès,[6] Delano and Pegis[7]).

By definition, the eigenvalues of the characteristic matrix $\mathbf{M}$ are the roots L of the equation (Kreyszig[4])

$$\det(\mathbf{M} - L\mathbf{I}) = 0$$

or

$$L^2 - (M_{11} + M_{22})L + \det(\mathbf{M}) = L^2 - (M_{11} + M_{22})L + 1 = 0$$

On the other hand, every matrix satisfies its own eigenvalue equations (Hamilton-Cayley equation, Lanczos[8])

$$\mathbf{M}^2 - (M_{11} + M_{22})\mathbf{M} + \mathbf{I} = 0 \tag{2.29}$$

In this equation $\mathbf{M}^2$ is expressed as a function of $\mathbf{M}$. If we multiply Eq. 2.29 by $\mathbf{M}$ and use Eq. 2.29 again to eliminate $\mathbf{M}^2$ we can express $\mathbf{M}^3$ as a function of $\mathbf{M}$. Setting $x = M_{11} + M_{22}$ we obtain

$$\mathbf{M}^3 = (x^2 - 1)\mathbf{M} - x\mathbf{I}$$

For $\mathbf{M}^4$ we obtain in the same fashion

$$\mathbf{M}^4 = (x^3 - 2x)\mathbf{M} - (x^2 - 1)\mathbf{I}$$

or in general

$$\mathbf{M}^p = [S_{p-1}(x)]\mathbf{M} - [S_{p-2}(x)]\mathbf{I} \tag{2.30}$$

S_{p-1} and S_{p-2} stand for Chebyshev polynomials of the order $p - 1$ and $p - 2$. They can be easily evaluated using the following recurrence relation (Abramowitz and Stegun[9]) (see Table 2.1 and Fig. 2.3):

$$S_p(x) - xS_{p-1}(x) + S_{p-2}(x) = 0 \tag{2.31}$$

The first two polynomials are $S_0(x) = 1$ and $S_1(x) = x$.

2.7. Transfer Matrix, Reflectance, and Transmittance

In Eqs. 2.20 and 2.21 we established a relationship between the electromagnetic wave functions E and H at the first ($z = z_1$) and the last

TABLE 2.1. Properties of the Chebyshev Polynomials S_m (x)

Definition: $S_{m-1}(x) = \dfrac{2 \sin [m \arccos (x/2)]}{\sqrt{4 - x^2}}$

$S_0(x) = 1$
$S_1(x) = x$
$S_2(x) = x^2 - 1$
$S_3(x) = x^3 - 2x$
$S_4(x) = x^4 - 3x^2 + 1$
$S_5(x) = x^5 - 4x^3 + 3x$
$S_6(x) = x^6 - 5x^4 + 6x^2 - 1$
$S_7(x) = x^7 - 6x^5 + 10x^3 - 4x$
$S_8(x) = x^8 - 7x^6 + 15x^4 - 10x^2 + 1$
$S_9(x) = x^9 - 8x^7 + 21x^5 - 20x^3 + 5x$
$S_{10}(x) = x^{10} - 9x^8 + 28x^6 - 35x^4 + 15x^2 - 1$
$S_{11}(x) = x^{11} - 10x^9 + 36x^7 - 56x^5 + 35x^3 - 6x$
$S_{12}(x) = x^{12} - 11x^{10} + 45x^8 - 84x^6 + 70x^4 - 21x^2 + 1$

Recurrence relations: $S_m(x) - xS_{m-1}(x) + S_{m-2}(x) = 0$

Coefficients: $S_m(x) = x^m - \begin{bmatrix} m-1 \\ 1 \end{bmatrix} x^{m-2} + \begin{bmatrix} m-2 \\ 2 \end{bmatrix} x^{m-4} \cdots$

$(z = z_{m+1})$ interface of a multilayer. Only inside the multilayer did we have to distinguish between positive- and negative-going waves. Consequently, no information about the massive media on both sides of the multilayer was necessary.

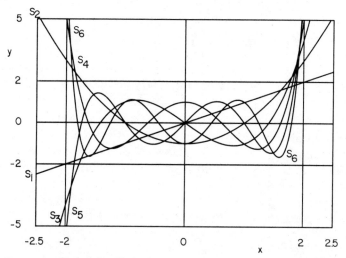

Figure 2.3. Plot of the Chebyshev polynomials $S_m(x)$ for $m = 1$ to 6 and $-2.5 < x < 2.5$.

This is different when we want to calculate the reflectance and transmittance of a multilayer relative to specific massive media.

In Eqs. 2.1 and 2.2, together with Eqs. 2.8 and 2.9, we used the following relationships between the electromagnetic wave functions E and H and E^+ and E^- (Berning[2]):

$$\begin{bmatrix} E \\ ZH \end{bmatrix} = \begin{bmatrix} 1 & 1 \\ n & -n \end{bmatrix} \begin{bmatrix} E^+ \\ E^- \end{bmatrix} \qquad (2.32)$$

Applying Eq. 2.32 to Eqs. 2.20 and 2.21 we obtain

$$\begin{bmatrix} E^+(z_1 - \delta) \\ E^-(z_1 - \delta) \end{bmatrix} = \begin{bmatrix} 1 & 1 \\ n & -n \end{bmatrix}^{-1} \begin{bmatrix} M_{11} & iM_{12} \\ iM_{21} & M_{22} \end{bmatrix} \begin{bmatrix} 1 & 1 \\ n & -n \end{bmatrix} \begin{bmatrix} E^+(z_{m+1} + \delta) \\ E^-(z_{m+1} + \delta) \end{bmatrix} \qquad (2.33)$$

or

$$\begin{bmatrix} E^+(z_1 - \delta) \\ E^-(z_1 - \delta) \end{bmatrix} = \begin{bmatrix} Q_{11} & Q_{12} \\ Q_{21} & Q_{22} \end{bmatrix} \begin{bmatrix} E^+(z_{m+1} + \delta) \\ E^-(z_{m+1} + \delta) \end{bmatrix} \qquad (2.34)$$

with

$$Q_{11} = \frac{M_{11} + iM_{21}/n_0 + in_sM_{12} + n_sM_{22}/n_0}{2}$$

$$Q_{12} = \frac{M_{11} + iM_{21}/n_0 - in_sM_{12} - n_sM_{22}/n_0}{2}$$

$$Q_{21} = \frac{M_{11} - iM_{21}/n_0 + in_sM_{12} - n_sM_{22}/n_0}{2}$$

$$Q_{22} = \frac{M_{11} - iM_{21}/n_0 - in_sM_{12} + n_sM_{22}/n_0}{2}$$

From microwave filter theory we adopt the name transfer matrix for this Q matrix (Young[10]).

In the exit medium ($z > z_{m+1}$) there is no negative-going (reflected) wave. We can consequently set $E^-(z_{m+1} + \delta) = 0$. This leads to

$$E^+(z_1 - \delta) = Q_{11}E^+(z_{m+1} + \delta)$$

$$E^-(z_1 - \delta) = Q_{21}E^+(z_{m+1} + \delta)$$

In as much as the

$$\text{Amplitude reflectance } \vec{R} = \frac{E^-(z_1 - \delta)}{E^+(z_1 - \delta)} = \frac{Q_{21}}{Q_{11}} \quad (2.35)$$

and the

$$\text{Amplitude transmittance } \vec{T} = \frac{E^+(z_{m+1} + \delta)}{E^+(z_1 - \delta)} = \frac{1}{Q_{11}} \quad (2.36)$$

we can derive out of Eq. 2.34

phase of reflectance

$$\vec{R} = \frac{n_0 M_{11} + in_0 n_s M_{12} - iM_{21} - n_s M_{22}}{n_0 M_{11} + in_0 n_s M_{12} + iM_{21} + n_s M_{22}} \quad (2.37)$$

and

phase of transmittance

$$\vec{T} = \frac{2n_0}{n_0 M_{11} + in_0 n_s M_{12} + iM_{21} + n_s M_{22}} \quad (2.38)$$

Amplitude reflectance $\vec{R}$ and transmittance $\vec{T}$ are connected with energy reflectance R and transmittance T through the Poynting vector. We have

reflectance   transmittance

$$R = \vec{R}\,\vec{R}^* \quad \text{and} \quad T = \frac{n_s}{n_0} \vec{T}\,\vec{T}^* = 1 - R \quad (2.39)$$

where $\vec{R}^*$ and $\vec{T}^*$ are the conjugate complex values of $\vec{R}$ and $\vec{T}$. We consequently can derive from Eqs. 2.37, 2.38, and 2.39

$$T = 1 - R = \frac{4n_0 n_S}{(n_0 M_{11} + n_s M_{22})^2 + (n_0 n_s M_{12} + M_{21})^2}$$

$$= \frac{4}{2 + (n_0/n_s) M_{11}^2 + n_0 n_s M_{12}^2 + M_{21}^2/n_0 n_s + (n_s/n_0) M_{22}^2} \quad (2.40)$$

2.8. Invariances of Reflectance and Transmittance

The elements of the characteristic matrix of a multilayer respond in a simple way to certain transformations of all indices of refraction of the individual layers of the multilayer.

1. *Multiply all indices with a constant factor.* The characteristic matrix of a single layer (Eq. 2.17) can be split into three matrices:

$$\begin{bmatrix} \cos\phi & \dfrac{i\sin\phi}{n} \\ in\sin\phi & \cos\phi \end{bmatrix} = \begin{bmatrix} 1 & 0 \\ 0 & n \end{bmatrix}\begin{bmatrix} \cos\phi & i\sin\phi \\ i\sin\phi & \cos\phi \end{bmatrix}\begin{bmatrix} 1 & 0 \\ 0 & \dfrac{1}{n} \end{bmatrix}$$

$$= \mathbf{S}(n)\mathbf{P}(\phi)\mathbf{S}\left(\frac{1}{n}\right) \qquad (2.41)$$

We observe that

$$\mathbf{S}\left(\frac{1}{n_1}\right)\mathbf{S}(n_2) = \mathbf{S}\left(\frac{n_2}{n_1}\right) \qquad (2.42)$$

Let us now go back to Eq. 2.21. Using Eqs. 2.41 and 2.42 we can write

$$\mathbf{M} = \mathbf{S}(n_1)\mathbf{P}(\phi_1)\mathbf{S}\left(\frac{n_2}{n_1}\right)\mathbf{P}(\phi_2)\,\mathbf{S}\left(\frac{n_3}{n_2}\right)\cdots\mathbf{S}\left(\frac{n_m}{n_{m-1}}\right)\mathbf{P}(\phi_m)\,\mathbf{S}\left(\frac{1}{n_m}\right)$$

or

$$\mathbf{M} = \mathbf{S}(n_1)\mathbf{F}(\phi_2,\,\phi_2,\,\ldots,\,n\text{-ratios})\mathbf{S}\left(\frac{1}{n_m}\right) \qquad (2.43)$$

Let us now replace all indices of refraction n by cn, premultiply Eq. 2.43 with $S(1/c)$, and postmultiply Eq. 2.43 with $S(c)$. We obtain

$$\begin{bmatrix} M_{11}(cn) & icM_{12}(cn) \\ \dfrac{iM_{21}(cn)}{c} & M_{22}(cn) \end{bmatrix} = \mathbf{S}(n_1)\mathbf{F}(\phi_1,\,\phi_2,\,\ldots,\,n\text{-ratios})\mathbf{S}\left(\frac{1}{n_m}\right)$$

$$= \begin{bmatrix} M_{11}(n) & iM_{12}(n) \\ iM_{21}(n) & M_{22}(n) \end{bmatrix}$$

or

$$M_{11}(cn) = M_{11}(n) \qquad cM_{12}(cn) = M_{12}(n)$$
$$\frac{M_{21}(cn)}{c} = M_{21}(n) \qquad M_{22}(cn) = M_{22}(n) \qquad (2.44)$$

2. *Replace all indices by reciprocal values.* Let us now define the matrix $\mathbf{S}$ as

$$\mathbf{S} = \begin{bmatrix} 0 & 1 \\ 1 & 0 \end{bmatrix}$$

and pre- and postmultiply the characteristic matrix of a single layer (Eq. 2.17) with it:

$$\begin{bmatrix} 0 & 1 \\ 1 & 0 \end{bmatrix} \begin{bmatrix} \cos\phi & \dfrac{i\sin\phi}{n} \\ in\sin\phi & \cos\phi \end{bmatrix} \begin{bmatrix} 0 & 1 \\ 1 & 0 \end{bmatrix} = \begin{bmatrix} \cos\phi & in\sin\phi \\ \dfrac{i\sin\phi}{n} & \cos\phi \end{bmatrix} \qquad (2.45)$$

The right side of Eq. 2.45 is the matrix of a single layer with n replaced by $1/n$. Observing that $\mathbf{SS} = \mathbf{I}$ we can transform Eq. 2.21 by pre- and postmultiplying with $\mathbf{S}$ into

$$\begin{bmatrix} M_{22}(n) & iM_{21}(n) \\ iM_{12}(n) & M_{11}(n) \end{bmatrix} = \mathbf{M}_1\left(\frac{1}{n_1}\right)\mathbf{M}_2\left(\frac{1}{n_2}\right) \cdots \mathbf{M}_m\left(\frac{1}{n_m}\right)$$

$$= \begin{bmatrix} M_{11}\left(\dfrac{1}{n}\right) & iM_{12}\left(\dfrac{1}{n}\right) \\ iM_{21}\left(\dfrac{1}{n}\right) & M_{22}\left(\dfrac{1}{n}\right) \end{bmatrix}$$

or

$$M_{11}\left(\frac{1}{n}\right) = M_{22}(n) \qquad M_{12}\left(\frac{1}{n}\right) = M_{21}(n)$$

$$M_{21}\left(\frac{1}{n}\right) = M_{12}(n) \qquad M_{22}\left(\frac{1}{n}\right) = M_{11}(n) \qquad (2.46)$$

Inserting these values together with the respective values for n_0 and n_s into Eqs. 2.37 and 2.38 we find

$$\vec{R}(n) = \vec{R}(cn) \qquad \text{(2.47)}$$

$$\vec{R}(n) = -\vec{R}\left(\frac{1}{n}\right) \qquad \text{(2.48)}$$

$$\vec{T}(n) = \vec{T}(cn) \qquad \text{(2.49)}$$

$$\vec{T}(n) = \vec{T}\left(\frac{1}{n}\right) \qquad \text{(2.50)}$$

or in words: The transmittance, reflectance, and phase of transmittance of a dielectric multilayer are *invariant* to multiplying *all* indices of refraction (including the indices of the surrounding medium and substrate) with a *constant factor* c or replacing all indices of refraction by their *reciprocal value*. The phase of reflectance is invariant only to constant factor changes. Replacing all indices by their reciprocal values results in a phase change of π.

TABLE 2.2   Systems with Equal Optical Performance

	System I	System II	System III
n_0	1.0	2.0	1.0
n_H	2.5	5.0	0.4
n_L	1.5	3.0	0.67
n_S	2.0	4.0	0.5

As examples three multilayer systems which exhibit the same transmittance, reflectance, phase of transmittance, and phase of reflectance ($+\pi$ for System III) are given in Table 2.2.

2.9. Split Filter Analysis

Let us consider two interfaces inside a multilayer, $z = z_k$ and $z = z_{k+1}$, bounding a spacer layer which has the refractive index n and the physical thickness d (Fig. 2.4). We may consider all the layers on top of $z = z_k$, including the interface, as System A. System A can be described by its complex reflection coefficient $\vec{r}_A$, as seen from inside the spacer, and its complex transmission coefficient $\vec{T}_A$. Similarly, System B consists of all the layers below $z = z_{k+1}$ and can be described, as seen from inside the spacer, by its coefficients $\vec{R}_B$ and $\vec{T}_B$.

In order to establish a relationship between the overall transmittance T and $\vec{r}_A$, $\vec{T}_A$, $\vec{R}_B$, and $\vec{T}_B$ we first like to expand our treatment of the reflectance and transmittance of a multilayer to include reverse reflectances.

With Eq. 2.32 we derived a relationship between the positive- and negative-going waves in the massive media on both sides of the multilayer and the total waves. We can treat the case when a wave is

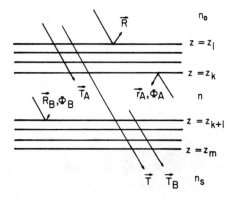

Figure 2.4.   Split filter analysis.

incident in the reverse direction by considering the massive medium $z > z_{m+1}$ as the incident medium and setting $E^+(z_1 - \delta) = 0$ (no reflected waves in the exit medium $z < z_1$). From Eq. 2.34

$$\begin{bmatrix} E^+(z_{m+1} + \delta) \\ E^-(z_{m+1} + \delta) \end{bmatrix} = \begin{bmatrix} Q_{11} & Q_{12} \\ Q_{21} & Q_{22} \end{bmatrix}^{-1} \begin{bmatrix} 0 \\ E^-(z_1 - \delta) \end{bmatrix}$$

$$= \begin{bmatrix} Q_{22} & -Q_{12} \\ -Q_{21} & Q_{11} \end{bmatrix} \begin{bmatrix} 0 \\ E^-(z_1 - \delta) \end{bmatrix}$$

which leads to the following formula for $\vec{r}$:

$$\vec{r} = \frac{E^+(z_{m+1} + \delta)}{E^-(z_{m+1} + \delta)} = -\frac{Q_{12}}{Q_{11}}$$

We can now write Eq. 2.34 in terms of $\vec{R}$, $\vec{r}$, and $\vec{T}$ (Berning[2])

$$\begin{bmatrix} E^+ (z_1 - \delta) \\ E^- (z_1 - \delta) \end{bmatrix} = \begin{bmatrix} \dfrac{1}{\vec{T}} & \dfrac{-\vec{r}}{\vec{T}} \\ \dfrac{\vec{R}}{\vec{T}} & \dfrac{(\vec{T}^2 - \vec{R}\vec{r})}{\vec{T}} \end{bmatrix} \begin{bmatrix} E^+ (z_{m+1} + \delta) \\ E^- (z_{m+1} + \delta) \end{bmatrix} \qquad (2.51)$$

since $Q_{11}Q_{22} - Q_{12}Q_{21} = 1$.

With Eq. 2.51 we can relate the positive- and negative-going waves in the surrounding medium and substrate of Fig. 2.4 to the amplitude reflectances and transmittances of Systems A and B and the phase thickness ϕ of the spacer. For this purpose we number the interfaces according to Fig. 2.4. The spacer layer is the kth layer being surrounded by interfaces at $z = z_k$ and z_{k+1}. The substrate starts after the mth layer at $z = z_{m+1}$.

$$\begin{bmatrix} E^+ (z_1 - \delta) \\ E^- (z_1 - \delta) \end{bmatrix} = \begin{bmatrix} \dfrac{1}{\vec{T}_A} & \dfrac{-\vec{r}_A}{\vec{T}_A} \\ \dfrac{\vec{R}_A}{\vec{T}_A} & \dfrac{(\vec{T}_A^2 - \vec{R}_A\vec{r}_A)}{\vec{T}_A} \end{bmatrix} \begin{bmatrix} e^{i\phi} & 0 \\ 0 & e^{-i\phi} \end{bmatrix}$$

$$\begin{bmatrix} \dfrac{1}{\vec{T}_B} & \dfrac{-\vec{r}_B}{\vec{T}_B} \\ \dfrac{\vec{R}_B}{\vec{T}_B} & \dfrac{(\vec{T}_B^2 - \vec{R}_B\vec{r}_B)}{\vec{T}_B} \end{bmatrix} \begin{bmatrix} E^+ (z_{m+1} + \delta) \\ E^- (z_{m+1} + \delta) \end{bmatrix} \qquad (2.52)$$

The second matrix on the right side of Eq. 2.52 transforms from the positive- and negative-going waves at $z = z_k + \delta$ to $z + z_{k+1} - \delta$.

If the massive medium at $z > z_{m+1}$ is the exit medium, then $E^-(z_{m+1} + \delta) = 0$ (no reflected wave in the exit medium). With this condition Eq. 2.52 reduces to

$$
\begin{bmatrix} E^+ (z_1 - \delta) \\ E^- (z_1 - \delta) \end{bmatrix} =
\begin{bmatrix} \dfrac{e^{i\phi}}{\vec{T}_A} & \dfrac{-\vec{r}_A e^{-i\phi}}{\vec{T}_A} \\ \dfrac{\vec{R}_A e^{i\phi}}{\vec{T}_A} & \dfrac{e^{-i\phi}(\vec{T}_A^2 - \vec{R}_A \vec{r}_A)}{\vec{T}_A} \end{bmatrix}
\begin{bmatrix} \dfrac{E^+ (z_{m+1} + \delta)}{\vec{T}_B} \\ \dfrac{\vec{R}_B E^+ (z_{m+1} + \delta)}{\vec{T}_B} \end{bmatrix}
$$

and leads to

$$
\vec{R} = \frac{E^-(z_1 - \delta)}{E^+(z_1 - \delta)} = \frac{\vec{R}_A + \vec{R}_B e^{-2i\phi}(\vec{T}_A^2 - \vec{r}_A \vec{R}_A)}{1 - \vec{r}_A \vec{R}_B e^{-2i\phi}} \tag{2.53}
$$

and

$$
\vec{T} = \frac{E^+(z_{m+1} - \delta)}{E^+(z_1 + \delta)} = \frac{\vec{T}_A \vec{T}_B e^{-i\phi}}{1 - \vec{r}_A \vec{R}_B e^{-2i\phi}} \tag{2.54}
$$

Setting $\vec{r}_A = |\vec{r}_A| e^{i\Phi_A}$, $\vec{R}_B = |\vec{R}_B| e^{i\Phi_B}$, $r = (|\vec{r}_A||\vec{R}_B|)^{1/2}$ and using Eq. 2.39 for $\vec{T}$ as well as for $\vec{T}_A$ and $\vec{T}_B$ we obtain from Eq. 2.51

$$
\frac{n_S T}{n_0} = \frac{(n T_A/n_0)(n_S T_B/n)}{\{1 - r \exp[-i(\Phi_A + \Phi_B - 2\phi)]\}\{1 - r \exp[i(\Phi_A + \Phi_B - 2\phi)]\}}
$$

and finally

$$
T = \frac{T_A T_B}{(1 - r)^2} \frac{1}{1 + [4r/(1 - r)^2]\sin^2[(\Phi_A + \Phi_B - 2\phi)/2]} \tag{2.55}
$$

Recapitulation of symbols:

$T \equiv$ transmittance through the complete multilayer

$T_A \equiv$ transmittance through System A

$T_B \equiv$ transmittance through System B

$r = \sqrt{r_A R_B}$ with r_A and R_B being the reflectances of Systems A and B for light incidence from the spacer

Φ_A, $\Phi_B \equiv$ phase changes on reflection associated with r_A and R_B

$\phi = 2\pi nd/\lambda$ with n, d being the refractive index and the physical thickness of the spacer layer

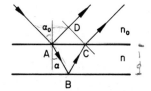

PHASE DIFFERENCE AT $\overline{DC}$

$= n(\overline{AB} + \overline{BC}) - n_0 \overline{AD}$

$= n(2d/\cos\alpha) - n_0(2d\tan\alpha\sin\alpha_0)$

$= 2nd(1 - n_0^2\sin^2\alpha_0/n^2)/\cos\alpha$

$= 2nd\cos^2\alpha/\cos\alpha = 2nd\cos\alpha$

Figure 2.5. Geometric explanation of why the effective optical thickness decreases with increasing angle of the incident light.

2.10. Nonnormal Light Incidence

Most thin film devices are applied at normal light incidence or small deviations from it. Since in Eqs. 2.5 and 2.16 the angle of incidence α appears in cosine functions we can assume normal incidence for all applications up to about $\alpha_0 = 20°$ (assuming $n_0 = 1$). When the angle of incidence is increased beyond $\alpha_0 = 20°$ dramatic changes occur. We will differentiate between p- and s-polarization of the incident light and deal with averages of very different optical behaviors. We will have to make two separate calculations, one for s-polarization and one for p-polarization.

On the other hand, calculations at an angle are the same as at normal incidence—only the thicknesses and indices are different. $\varphi = \frac{4\pi}{\lambda} nd \cos\alpha$

Let us now look at Eq. 2.5. The fact that the effective optical thickness $d = \frac{7}{n_m} - 7$, should decrease with increasing angle of incidence appears as a paradox at first. Is not the path length traveled through the film at an angle longer? The resolution of the paradox lies in the fact that the phase thickness ϕ is proportional to the path difference between a film and no film (Fig. 2.5) [$n_0 \times AD$ more than compensates for the increase in $n \times (AB + BC)$].

Equation 2.5 also explains the fact that the transmittance and reflectance curves of optical interference coatings shift to shorter wavelengths with increasing angle of the incident light (Fig. 9.1). A coating at higher angle of incidence corresponds to a coating with thinner layers. This is a very characteristic behavior of optical interference coatings and can be used to differentiate them from absorption-type filters which shift (albeit very little) in the opposite direction (longer wavelengths) (Schott[11]).

We see that, with increasing angle of incidence, differences between refractive indices (a high and a low index, for example) increase in the

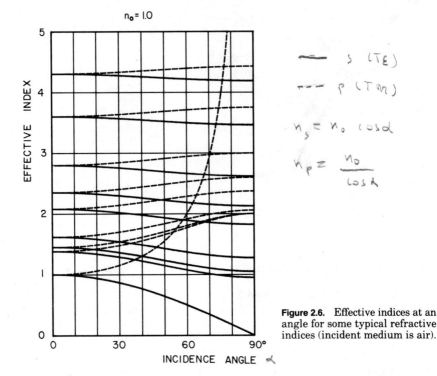

Figure 2.6. Effective indices at an angle for some typical refractive indices (incident medium is air).

s-plane and decrease in the p-plane while the absolute values decrease in the s-plane and increase in the p-plane.

In Fig. 2.6 we show the effective indices (Eq. 2.16) as a function of the incidence angle for the incidence medium air. We observe that the higher the index the lower the polarization split. The practical consequence is that designs with higher refractive indices are less angle sensitive.

If the incidence medium is no longer air (cemented filter) (Fig. 2.7) an exaggeration of the tendencies of Fig. 2.6 occurs. It can even happen that two materials with different index of refraction have the same effective index at a certain angle (crossings of dashed curves in Fig. 2.7). This fact can be used to design polarizers (MacNeille[12]).

2.11. Description of Designs

A dielectric multilayer design is fully described by the (real) refractive indices and thicknesses of its individual layers and the refractive indices of the two bounding massive media (incidence and exit medium or surrounding medium and substrate). The angle of incidence α_0 and the wavelength λ specify how and where it is used.

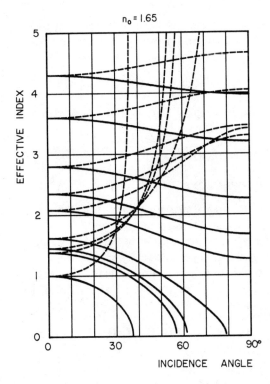

$n_0 = 1.65$

EFFECTIVE INDEX

INCIDENCE   ANGLE

Figure 2.7. Effective indices at an angle of some typical refractive indices (incident medium is glass with index 1.65).

For the thickness description the optical thickness is generally preferred, but not in its usual definition nd but as quarter-wave optical thickness (QWOT), since quarter-wave films play such a dominant role in thin film optics. $QWOT$ is defined as the wavelength at which a film is 0.25 wavelength thick, or, with Eq. 2.5,

$$\phi = \frac{\pi}{2} \frac{QWOT}{\lambda} \quad \text{or} \quad QWOT = 4nd \cos \alpha \quad (2.56)$$

How do we now write the formula for the design? The indices of refraction of the massive media appear in front (incident medium, most of the time the surrounding medium) and behind (exit medium, most of the time the substrate). The description of the actual multilayer, separated by vertical lines, is placed in-between. Capital letters stand for layers 0.25 wavelength thick at a design wavelength λ_0. Individual or groups of layers with different thicknesses are assigned a

$$\text{Fractional thickness} = \frac{QWOT}{\lambda_0} \quad (2.57)$$

[e.g., H/2, 1.05(ABABABBA)]. Periodic multilayers (Sec. 2.6) are described by placing the layers making up the base period (often called elements) between parentheses and indicating the number of repetitions by a power:

$$ABABABAB = (AB)^4$$

With this convention, a particular design can be described in many different ways. For example, the design

$$1.0 \,|\, (AB)^6 \; 1.1A \; 1.2(BA)^6 \,|\, 1.52$$

can also be written in the following way:

$$1.0 \,|\, A/2 \; (A/2 \; B \; A/2)^6 \; 1.2((A/2 \; B \; A/2)^6 \; A/2) \,|\, 1.52$$

This second description is of great theoretical value because the basic building blocks of the design are now symmetrical periods (Sec. 3.1).

Designs with half-wave-thick layers invite different theoretical treatment depending on the way they are written. For example, the design

$$1.52 \,|\, HLHLHHLHLH \; L \; HLHLHHLHLH \; L \; HLHLHHLHLH \,|\, 1.52$$

can also be written as

$$1.52 \,|\, (HL)^2 \; H \; (HL)^5 \; H \; (HL)^5 \; H \; (HL)^2 \; H \,|\, 1.52$$

or as

$$1.52 \,|\, (HLHLH) \; (HLHLHLHLHLH)^2 \; (HLHLH) \,|\, 1.52$$

In addition to the incidence angle α_0 the so-called match angle β is often used. This angle becomes important when a design is intended and matched for one angle of incidence (β) but analyzed at another (α). With α, β, and Eq. 2.56, Eq. 2.5 becomes

$$\phi = \frac{\pi}{2} \frac{QWOT_\beta}{\lambda} \frac{\sqrt{1 - (n_0 \sin \alpha/n)^2}}{\sqrt{1 - (n_0 \sin \beta/n)^2}} \tag{2.58}$$

The reflectance and transmittance are best given as functions of the relative wavenumber λ_0/λ. This way the symmetries of the sine and cosine functions of the characteristic matrix (Eq. 2.17) are preserved. The use of QWOT as a thickness unit and λ_0/λ as an independent variable for reflectance and transmittance plots often causes some conceptual problems. When a design is developed to a wavelength specification rather than to a wavenumber (or frequency) specification one has to remember that a long-wavelength-pass filter is a short-fre-

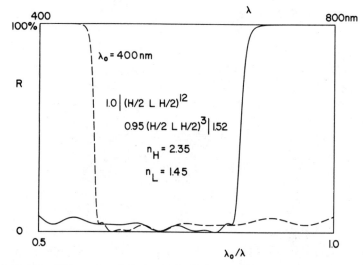

Figure 2.8. The reflectance of a design by Thelen[13] as a function of wave-number (solid curve) and wavelength (dashed curve) (λ_0 = 400 nm).

quency-pass filter and that, relative to the wavenumber presentation, features at shorter wavelengths are compressed and at longer wavelengths stretched (Fig. 2.8). If one wants to shift a design on a wavelength scale to a longer wavelength by 20 percent one has to multiply all thicknesses by 1.2 while on a wavenumber scale this multiplication of all thicknesses by 1.2 causes a shift by 20 percent to shorter wavenumbers (Fig. 2-9).

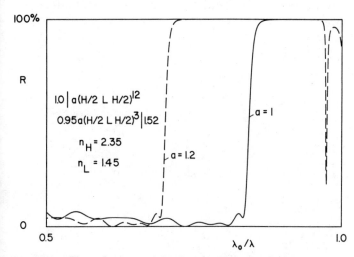

Figure 2.9. The reflectance of the design of Fig. 2.8 (solid curve) and for 20 percent increased optical thickness of all layers (dashed curve).

2.12.   Problems and Solutions

Problem 2.1

Calculate the reflectance of a thin film with index n on a substrate with index n_s at the quarter-wave position. The index of the incident medium is n_0.

Solution.   Using Eqs. 2.18 and 2.40 we obtain

$$R = 1 - \frac{4n_0 n_s n^2}{(n_0 n_s + n^2)^2} = \frac{(n_0 n_s - n^2)^2}{(n_0 n_s + n^2)^2} \tag{2.59}$$

Problem 2.2

What quarter-wave optical thickness should a thin film have at normal light incidence ($\text{QWOT}_{\alpha=0}$) when it has a specified $\text{QWOT}_{\alpha=\beta}$ at a given nonnormal incidence angle β (match angle)?

Solution.   Using Eq. 2.58

$$\text{QWOT}_{\alpha=0} = \text{QWOT}_{\alpha=\beta} \frac{1}{\sqrt{1 - (n_0 \sin \beta / n)^2}} \tag{2.60}$$

Problem 2.3

Calculate the reflectance and transmittance of a stack of half-wave-thick layers between a medium (n_0) and substrate (n_s).

Solution.   Using Eq. 2.19

$$\mathbf{M} = \begin{bmatrix} \pm 1 & 0 \\ 0 & \pm 1 \end{bmatrix}$$

Using Eq. 2.40

$$T = \frac{4}{2 + n_0/n_s + n_s/n_0} = \frac{4}{(n_0 + n_s)^2/n_0 n_s} = \frac{4n_0 n_s}{(n_0 + n_s)^2}$$

and

$$R = 1 - T = \left(\frac{n_0 - n_s}{n_0 + n_s}\right)^2 \tag{2.61}$$

These relations are equivalent to the formulas for the transmittance and reflectance of the uncoated substrate, or a stack of half-wave-thick films does not alter the reflectance of the bare substrate at the half-wave point (Sec. 3.5).

Problem 2.4

Calculate the transmittance of a stack of p quarter-wave-thick layers with alternating refractive indices n_1 and n_2 between a medium (n_0) and substrate (n_s). Differentiate between the cases of p even and odd.

Solution. Using Eq. 2.18 for p even:

$$p = 2: \qquad \mathbf{M} = \begin{bmatrix} \dfrac{-n_2}{n_1} & 0 \\ 0 & \dfrac{-n_1}{n_2} \end{bmatrix}$$

$$p = 4: \qquad \mathbf{M} = \begin{bmatrix} \left(\dfrac{n_2}{n_1}\right)^2 & 0 \\ 0 & \left(\dfrac{n_1}{n_2}\right)^2 \end{bmatrix}$$

$$p = p: \qquad \mathbf{M} = \begin{bmatrix} \left(\dfrac{-n_2}{n_1}\right)^{p/2} & 0 \\ 0 & \left(\dfrac{-n_1}{n_2}\right)^{p/2} \end{bmatrix}$$

with Eq. 2.40

$$T = \frac{4}{2 + (n_0/n_S)(n_2/n_1)^p + (n_s/n_0)(n_1/n_2)^p}$$

When $n_1 > n_2$ and p large, then

$$T = 4\left(\frac{n_0}{n_S}\right)\left(\frac{n_2}{n_1}\right)^p \qquad (2.62)$$

For p odd:

$$\mathbf{M} = \begin{bmatrix} 0 & \dfrac{i(-n_2/n_1)^{(p-1)/2}}{n_1} \\ in_1\left(\dfrac{-n_1}{n_2}\right)^{(p-1)/2} & 0 \end{bmatrix}$$

$$T = \frac{4}{2 + \dfrac{n_0 n_s}{n_1^2}\left(\dfrac{n_2}{n_1}\right)^{p-1} + \dfrac{n_1^2}{n_0 n_s}\left(\dfrac{n_1}{n_2}\right)^{p-1}}$$

When $n_1 > n_2$ and p large, then

$$T = \frac{4n_0 n_s (n_2/n_1)^{p-1}}{n_1^2} \tag{2.63}$$

Problem 2.5

Find an approximation for the characteristic matrix of a half-wave-thick or quarter-wave-thick layer in the vicinity of the half-wave or quarter-wave positions.

Solution. When $\phi = 90° \pm \delta$: $\sin \phi = 1$ and $\cos \phi = \mp\delta$. Then, with Eq. 2.17,

$$\mathbf{M} = \begin{bmatrix} \mp\delta & \dfrac{i}{n} \\ in & \mp\delta \end{bmatrix} \tag{2.64}$$

When $\phi = 180° \pm \delta$: $\sin \phi = \mp\delta$ and $\cos \phi = -1$. Then

$$\mathbf{M} = \begin{bmatrix} -1 & \dfrac{\mp i\delta}{n} \\ \mp in\delta & -1 \end{bmatrix} \tag{2.65}$$

In both cases $1/T$ becomes an entire rational function of δ

$$\frac{1}{T} = a_0 \delta^m + a_1 \delta^{m-1} + \cdots + a_m$$

Problem 2.6

Write a computer program to calculate the reflectance of an arbitrary dielectric multilayer design using the results of Secs. 2.5 and 2.6.

Solution. The design is characterized by the following numbers:

1. The index of refraction of the surrounding medium
2. A set of three numbers in proper sequence for each element of the design:
   a. An identifying number indicating to which period the layer belongs
   b. An index of refraction
   c. A quarter-wave optical thickness

3. A set of three numbers for each period in proper sequence (starting from the surrounding medium):
   a. An identifying number matching in value one of the numbers entered under 2.a. and indicating by its sign whether the period is to be supplemented symmetrically ($-$) or not ($+$)
   b. A shift factor which is applied to the optical thicknesses of all layers of the period
   c. A power indicating how often the period is to be repeated
4. The index of the substrate
5. Incidence and match angle
6. Wavenumber sweep

The program consists of the following parts:

1. Input/edit
2. Verify/print input
3. Precalculate reoccurring numbers
4. Initialize and start polarization loop
5. Initialize and start wavenumber loop
6. Initialize and start total matrix loop
7. Initialize, start, and complete period matrix loop, apply proper shift, test for symmetry, and, if symmetric, supplement
8. Calculate Chebyshev polynomials, raise period matrices to proper power, and multiply into total matrix
9. Calculate reflectance
10. Complete wavenumber loop
11. Complete polarization loop

Program Listing for Problem 2.6

```
100   ! PROGRAM CALCULATES REFLECTANCE OF A NONABSORBING
101   ! MULTILAYER AT NONNORMAL INCIDENCE AS A FUNCTION
102   ! OF WAVELENGTH
103   OPTION BASE 1
104   DIM Id1(30),Qwot(30),Refr_ind(30)
105   DIM Id2(30),Fract_th(30),N_periods(30)
106   DIM Cos_inc(30),Cos_match(30)          double
107   DEG
108   !
109   ! INPUT DATA
110   !
111 Dsgndata:DATA 1,1.52         ! INDEX INC. MED, INDEX SUB-
112                              ! STRATE
113   DATA 1,400,1.45            ! ID, QWOT, INDEX OF PERIOD
114                              ! ELEMENTS
115   DATA 1,400,2.35
116   DATA 2,800,1.38
117   DATA 2,1600,2.1
118   DATA 2,800,1.63
119   DATA 0,0,0                 ! END OF ELEMENT INPUT
120   DATA -1,1.07,2             ! ID, FRACT. THICKN., NUMBER
121                              ! OF PERIODS, NEG. ID MEANS
122                              ! SYMMETRIC SUPPLEMENT
123   DATA -1,1,11
124   DATA -1,1.07,2
125   DATA 2,.625,1
126   DATA 0,0,0                 ! END OF PERIOD STRUCTURE IPT
127   DATA 400,50,800            ! WAVENUMBER SWEEP
128   DATA 45,45                 ! INCIDENCE ANGLE, MATCH ANGLE
129   !
130   ! READ DATA
131   !
132   RESTORE Dsgndata
133   READ Refr_ind0,Refr_ind9 ! INDEX INC. MEDIUM, SUBSTRATE
134   FOR I=1 TO 30                        @@
135      READ Id1(I),Qwot(I),Refr_ind(I)
136      IF Id1(I)=0 THEN 140            while (1.82f)
137      Tot_el=I                   ! Tot_el: TOTAL NUMBER OF ELE-
138                                 ! MENTS
139   NEXT I
140   FOR I=1 TO 30
141      READ Id2(I),Fract_th(I),N_periods(I)
142      IF Id2(I)=0 THEN 146
143      Tot_per=I                  ! Tot_per: TOTAL NUMBER OF
```

store in file

Program Listing for Problem 2.6 (Continued)

```
144                          ! PERIODS
145    NEXT I
146    READ W_start,W_incr,W_final
147    READ Inc_angle,Match_angle
148    !
149    ! LIST INPUT
150    !
151    PRINT
152    PRINT "n(med)=";Refr_ind0
153    PRINT "ID, QWOT ,n"
154    FOR I=1 TO Tot_el
155       PRINT Id1(I),Qwot(I),Refr_ind(I)
156    NEXT I
157    PRINT "ID, FRACT.THICKN, POWER"
158    FOR I=1 TO Tot_per
159       PRINT Id2(I),Fract_th(I),N_periods(I)
160    NEXT I
161    PRINT "n(sub)=";Refr_ind9
162    PRINT "INCIDENCE/MATCH ANGLE:";Inc_angle;Match_angle
163    PRINT "WAVENUMBER SWEEP:";W_start;W_incr;W_final
164    PRINT
165    PRINT "S-PLANE"
166    !
167    ! CALCULATIONS
168    !
169    I=1
170    Cos_inc(I)=SQR(1-(Refr_ind0*SIN(Inc_angle)/Refr_ind(I))^2)
171    Partial=SQR(1-(Refr_ind0*SIN(Match_angle)/Refr_ind(I))^2)
172    Cos_match(I)=Cos_inc(I)/Partial
173    I=I+1
174    IF I<=Tot_el THEN 170
175    Cos_sub=SQR(1-(Refr_ind0*SIN(Inc_angle)/Refr_ind9)^2)
176    Cos_med=COS(Inc_angle)
177    !
178    ! POLARIZATION LOOP
179    !
180    Pol=1                 ! Pol=1 MEANS S-POLARIZATION,
181    !                       Pol=0 MEANS P-POLARIZATION
182    !
183    ! WAVELENGTH LOOP
184    !
185    W_run=W_start
186    IF Pol=0 THEN 190
187    N1=Refr_ind0*Cos_med
```

(handwritten annotations: "do ... while" near line 167; "double polarization ( )" near line 178)

Program Listing for Problem 2.6 *(Continued)*

```
188   N8=Refr_ind9*Cos_sub
189   GOTO 195
190   N1=Refr_ind0/Cos_med
191   N8=Refr_ind9/Cos_sub
192   !
193   ! FINAL MATRIX LOOP
194   !
195   F11=1  ! FINAL CHARACTERISTIC MATRIX:  F11   iF12
196   !                                      iF21   F22
197   F12=0
198   F21=0
199   F22=1
200   I=1
201   !
202   ! PERIOD MATRIX:   E11   iE12          period mat( )
203   !                  iE21   E22
204   E11=1
205   E12=0
206   E21=0
207   E22=1
208   J=1
209   Sym_pointr=0        ! Sym_pointr IS POINTER FOR SYMMETRY
210   IF ABS(Id2(I))<>ABS(Id1(J)) THEN 226
211   Phi=Qwot(J)*Cos_match(J)*Fract_th(I)*90/W_run
212   H11=COS(Phi)
213   X1=SIN(Phi)
214   IF Pol=0 THEN 217
215   X2=Refr_ind(J)*Cos_inc(J)
216   GOTO 218
217   X2=Refr_ind(J)/Cos_inc(J)
218   H12=X1/X2
219   H21=X1*X2
220   X1=E11*H11-E12*H21
221   E12=E11*H12+E12*H11
222   X2=E21*H11+E22*H21
223   E22=E22*H11-E21*H12
224   E11=X1
225   E21=X2                                 do ... while
226   J=J+1
227   IF J<=Tot_el THEN 210      ! END OF PERIOD MATRIX LOOP BEFORE
228                              ! CONSIDERING SYMMETRY AND POWER
229   IF Id2(I)<0 THEN Sym_pointr=1  ! SET POINTER IF PERIOD IS
230                                  ! HALFSTRUCTURE
231   IF Sym_pointr<>1 THEN 241      ! IF HALFSTRUCTURE, MULTIPLY
```

main {

call chebyshev( )

Program Listing for Problem 2.6 (Continued)

```
232                                    ! WITH REVERSE MATRIX    reverse-mat( )
233    X1=E11*E22-E12*E21
234    E12=2*E11*E12
235    E21=2*E21*E22
236    E22=X1
237    E11=X1                  ! END OF HALFSTRUCTURE SUPPLEMENT
238    !
239    ! CHEBYSHEV POLYNOMIALS     chebyshev( )
240    !
241    X1=E11+E22
242    X7=0
243    X5=1
244    X6=N_periods(I)
245    IF X6=1 THEN 251
246    X4=X1*X5-X7
247    X7=X5
248    X5=X4
249    X6=X6-1
250    GOTO 245          ! END OF CALCULATION OF CHEBYSHEV POLS
251    X11=E11*X5-X7     ! RAISE PERIOD MATRIX TO PROPER POWER
252                      ! AND MULTIPLY WITH TOTAL MATRIX
253    X12=E12*X5
254    X21=E21*X5
255    X22=E22*X5-X7
256    X7=F11*X11-F12*X21
257    F12=F11*X12+F12*X22
258    X5=F21*X11+F22*X21
259    F22=F22*X22-F21*X12
260    F11=X7
261    F21=X5
262    I=I+1                    Call per_mat( )
263    IF I<=Tot_per THEN 204   ! END OF FINAL MATRIX LOOP
264    !
265    ! TRANSMITTANCE
266    !
267    X1=(N1*F11/N8+F22)^2+(N1*F12+F21/N8)^2
268    Trans=4*N1/(N8*X1)
269    Refl=100*(1-Trans)
270    PRINT W_run;Refl
271    IF W_run<W_final THEN     ! END OF WAVENUMBER LOOP
272       W_run=W_run+W_incr
273       GOTO 195               Call to f^n  char_mat( )
274    END IF                          per_mat( )
275    IF Pol=1 AND Inc_angle<>0 THEN ! END OF POLARIZATION LOOP
```

chebyshev( )

Program Listing for Problem 2.6 (Continued)

```
276    Pol=0
277    PRINT
278    PRINT "P-PLANE"
279    GOTO 185
280  END IF
281  END
```

Sample Calculation for Problem 2.6

```
n(med)= 1
ID, QWOT ,n
  1        400       1.45
  1        400       2.35
  2        800       1.38
  2       1600       2.1
  2        800       1.63
ID, FRACT.THICKN, POWER
 -1        1.07      2
 -1        1         11
 -1        1.07      2
  2         .625     1
n(sub)= 1.52
INCIDENCE/MATCH ANGLE: 45  45
WAVENUMBER SWEEP: 400  50  800


S-PLANE
 400   6.2391206904
 450    .711926657774
 500   2.26195897723
 550    .533536319773
 600   1.36785784176
 650   9.25908034865
 700   99.9356196818
 750   99.9998996677
 800   99.9999890969


P-PLANE
 400    .650476879505
 450    .216812003614
 500    .0170857334514
 550    .196855378088
 600   1.50952759634
```

Sample Calculation for Problem 2.6 (*Continued*)

```
650   1.27959045793
700   1.88337357437
750   99.9645425643
800   99.9972098949
```

Problem 2.7

With the transformation $n \rightarrow n + ik$ the formulas for the characteristic matrix, reflectance, transmittance, etc., can be extended to absorbing films. Write a computer program for the calculation of a sequence of absorbing films on a nonabsorbing substrate at normal light incidence.

Solution

Program Listing for Problem 2.7

```
100    ! PROGRAM CALCULATES TRANSMITTANCE (Trans) AND THE
101    ! REFLECTANCE FROM BOTH SIDES (Ref0 AND Ref9) OF A
102    ! STACK OF ABSORBING FILMS ON A NONABSORBING SUB-
103    ! STRATE - n AND k ARE ASSUMED TO HAVE NO DISPERSION -
104    ! NORMAL LIGHT INCIDENCE
105    OPTION BASE 0
106    RAD
107    DIM Fn(30),Fk(30),Physthickn(30)
108    !
109    ! INPUT/EDIT"
110    !
111 Dsgndata:DATA 1,1.52      ! INDICES OF INCIDENT MEDIUM
112                           ! AND SUBSTRATE
113    DATA 1.46,0,75         ! n, k, d (PHYSICAL THICKNESS)
114                           ! OF FIRST FILM
115    DATA 1.74,2.96,4
116    DATA 0,0,0             ! END OF INPUT OF FILM DATA
117    DATA 400,50,700        ! WAVELENGTH SWEEP
118    !
119    ! READ IN DATA AND LIST INPUT
120    !
121    PRINT
122    PRINT
123    PRINT "ABSORBING LAYERS"
124    PRINT
125    RESTORE Dsgndata
126    READ N0,N9
```

Program Listing for Problem 2.7 (*Continued*)

```
127    PRINT "N0=";N0;"NS=";N9
128    PRINT "   n        k          d"
129    N_layers=1
130    FOR I=1 TO 30
131       READ Fn(I),Fk(I),Physthickn(I)
132       PRINT Fn(I),Fk(I),Physthickn(I)
133       IF Physthickn(I)=0 THEN Wave_ipt
134       N_layers=I
135       NEXT I
136 Wave_ipt:  READ W_start,W_incr,W_final
137    PRINT
138    !
139    ! START CALCULATIONS
140    !
141    PRINT " Wavelgth    Trans           Ref0              Ref9"
142 Startcalc: W_run=W_start    ! W_run IS CURRENT WAVELENGTH
143 Startpnt: A11=1             ! STARTING MATRIX   1+i0      0+i0
144                             !                   0+i0      1+i0
145    A12=0
146    A21=0
147    A22=1
148    B11=0
149    B12=0
150    B21=0
151    B22=0
152    FOR I=1 TO N_layers
153    !
154    ! MATRIX OF CURRENT FILM:    C11+iD12      C12+iD12
155    !                            C21+iD21      C22+iD22
156    !
157    Alfa=2*PI*Physthickn(I)*Fn(I)/W_run
158    Beta=Alfa*Fk(I)/Fn(I)
159    Ct=COS(Alfa)
160    St=SIN(Alfa)
161    Sh=(EXP(Beta)-EXP(-Beta))/2
162    Ch=(EXP(Beta)+EXP(-Beta))/2
163    C11=Ct*Ch
164    D11=St*Sh
165    C12=(Fn(I)*Ct*Sh-Fk(I)*St*Ch)/(Fn(I)*Fn(I)+Fk(I)*Fk(I))
166    D12=(Fn(I)*St*Ch+Fk(I)*Ct*Sh)/(Fn(I)*Fn(I)+Fk(I)*Fk(I))
167    C21=(Fn(I)*Ct*Sh+Fk(I)*St*Ch)
168    D21=(Fn(I)*St*Ch-Fk(I)*Ct*Sh)
169    !
170    ! MATRIX OF MULTILAYER:      E11+iF11      E12+iF12
```

Program Listing for Problem 2.7 (*Continued*)

```
171   !                           E21+iF21     E22+iF22
172   !
173   E11=A11*C11-B11*D11+A12*C21-B12*D21
174   F11=B11*C11+A11*D11+B12*C21+A12*D21
175   E12=A11*C12-B11*D12+A12*C11-B12*D11
176   F12=B11*C12+A11*D12+B12*C11+A12*D11
177   E21=A21*C11-B21*D11+A22*C21-B22*D21
178   F21=B21*C11+A21*D11+B22*C21+A22*D21
179   E22=A21*C12-B21*D12+A22*C11-B22*D11
180   F22=B21*C12+A21*D12+B22*C11+A22*D11
181   A11=E11
182   B11=F11
183   A12=E12
184   B12=F12
185   A21=E21
186   B21=F21
187   A22=E22
188   B22=F22
189   NEXT I                      ! END OF LAYER LOOP
190   !
191   ! CALCULATE Trans, Ref0, AND Ref9
192   !
193   A11=N0*E11+N0*N9*E12
194   A12=E21+N9*E22
195   A21=N0*F11+N0*N9*F12
196   A22=F21+N9*F22
197   B11=N9*E22+N0*N9*E12
198   B12=E21+N0*E11
199   B21=N9*F22+N0*N9*F12
200   B22=F21+N0*F11
201   Trans=400*N0*N9/((A11+A12)^2+(A21+A22)^2)
202   Partial=100*((A11-A12)^2+(A21-A22)^2)
203   Ref0=Partial/((A11+A12)^2+(A21+A22)^2)
204   Partial=100*((B11-B12)^2+(B21-B22)^2)
205   Ref9=Partial/((B11+B12)^2+(B21+B22)^2)
206   PRINT W_run;TAB(12),Trans;TAB(28),Ref0;TAB(45),Ref9
207   !
208   ! END OF WAVELENGTH LOOP
209   !
210   IF W_run<W_final THEN
211      W_run=W_run+W_incr
212      GOTO Startpnt
213      END IF
214   END
```

Sample Calculation for Problem 2.7

ABSORBING LAYERS

NO= 1 NS= 1.52

n	k	d
1.46	0	75
1.74	2.96	4
0	0	0

Wavelgth	Trans	Ref0	Ref9
400	67.4195745177	3.54756453183	10.9826373569
450	71.5866452393	1.07778391948	8.90890594789
500	74.354207949	.136602059253	6.93744896358
550	76.2304866531	.0241775925538	5.30194264206
600	77.5638200242	.310047966073	4.03054916129
650	78.5679346568	.758828017825	3.07587034523
700	79.3687136002	1.25037366276	2.37405173187

3

General Design Methods

We will be discussing two types of design methods: (1) Those which start with thin films available in practice, combine them into building blocks, and then use physical or mathematical models to accomplish the final design and (2) those which start with a theoretical concept (e.g., equal ripple performance), translate it into a multilayer constructed of thin films generally not available in practice, and then try to find ways to replacing them with a combination of films available in practice.

Both ways have their advantages and disadvantages. The first approach generally leads to produceable designs sooner, adapts to physical limitations of thin films better, and is more responsive to new applications and practical problems. The second approach generally leads to ultimately better designs, more often finds new multilayer configurations, and is more challenging to the designer.

3.1. Equivalent Layers

With Eq. 2.17 we introduced the characteristic matrix of a single layer and with Eqs. 2.20, 2.21, and 2.22 the characteristic matrix of a sequence of layers. They both are 2 by 2 matrices with real elements in the principal diagonal and purely imaginary elements elsewhere, yet they differ in so far as the two elements of the principal diagonal in the case of the single layer equal each other while the corresponding elements of the multilayer generally do not. However, from Sec. 2.5 we know that the two elements in the principal diagonal of the characteristic matrix of a sequence of layers are equal ($M_{11} = M_{22}$) whenever the multilayer is symmetric. Since the characteristic matrix optically describes a sequence of layers fully we can conclude that a sequence of layers is equivalent to a single film whenever it is symmetric (Herpin[1]). The single film is called the *equivalent layer* of the sequence.

As a consequence we should be able to assign an *equivalent index* of refraction N and an *equivalent thickness* Γ to a symmetric sequence of layers. Comparing Eq. 2.17 with Eqs. 2.20, 2.21, and 2.22 we arrive at the following definitions (Epstein[2]):

$$\text{Equivalent index } N = +\sqrt{\frac{M_{21}}{M_{12}}} \qquad (3.1)$$

and

$$\text{Equivalent thickness } \Gamma = \arccos M_{11} \qquad (3.2)$$

It is important to remember that we established the equivalence mathematically and not physically. Consequently, since the elements of the characteristic matrix $\mathbf{M}$, M_{11}, M_{12}, M_{21}, and M_{22}, depend on the wavenumber, polarization, and angle of incidence, both N and Γ depend very much on the same parameters and have to be recomputed whenever a change in these parameters occurs.

3.1.1. Equivalent layers in periodic multilayers

Depending on the signs of M_{12} and M_{21}, we can distinguish between two cases:

1. M_{12} and M_{21} have equal signs ($+$ or $-$). Then

$$N = \sqrt{\frac{M_{12}}{M_{21}}} = \text{real}$$

Also, since $M_{11}M_{22} + M_{12}M_{21} = 1$ (Eq. 2.23) and $M_{11} = M_{22}$

$$\cos \Gamma = M_{11} = \sqrt{1 - |M_{12}M_{21}|} < 1$$

As a consequence, $\cos \Gamma$ and $\sin \Gamma$ are real trigonometric functions.

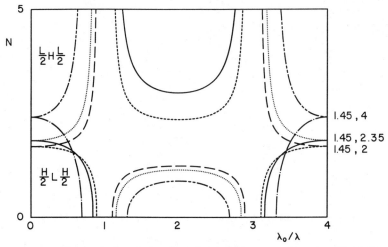

Figure 3.1. The equivalent index N of the three-layer combinations H/2 L H/2 and L/2 H L/2 with $n_L = 1.45$ and $n_H = 2$, 2.35, and 4. The central part of the dash-dotted curve is out of range.

2. M_{12} and M_{21} have unequal signs. Then

$$N = \sqrt{\frac{M_{12}}{M_{21}}} = \text{imaginary}$$

and

$$\cos \Gamma = M_{11} = \sqrt{1 + |M_{12}M_{21}|} > 1$$

Now $\cos \Gamma$ and $\sin \Gamma$ are no longer real trigonometric functions, they become the *hyperbolic functions* $\cosh \Gamma$ and $\sinh \Gamma$.

As examples we show in Fig. 3.1 the equivalent index N and in Fig. 3.2 the equivalent thicknesses Γ for the three-layer combinations H/2 L H/2 and L/2 H L/2 with $n_H = 2.0$, 2.35, and 4 and $n_L = 1.45$. Values are given only in the regions where N is real.

The distinction between cases (1) and (2) are of particular importance for periodic multilayers composed of symmetric periods. Let us construct a periodic multilayer by repeating a basic sequence ABA p times:

$$(\text{ABA})_1 \ (\text{ABA})_2 \ (\text{ABA})_3 \ \cdots \ (\text{ABA})_p$$

In terms of the equivalent index N_{ABA} and equivalent thickness Γ_{ABA}, the characteristic matrix of the periodic multilayer takes the following form, depending on whether N_{ABA} is real or not:

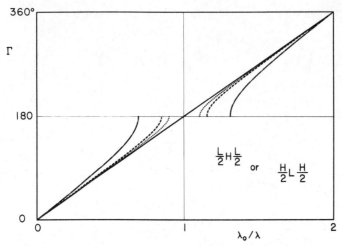

Figure 3.2. The equivalent thickness Γ of the three-layer combinations H/2 L H/2 and L/2 H L/2 with $n_L = 1.45$ and $n_H = 2$, 2.35, and 4. The larger is n_H/n_L, the wider the gap.

1. N_{ABA} is *real*:

$$M = \begin{bmatrix} \cos p\Gamma_{ABA} & \dfrac{i \sin p\Gamma_{ABA}}{N_{ABA}} \\ iN_{ABA} \sin p_{ABA} & \cos p_{ABA} \end{bmatrix} \qquad (3.3)$$

If we increase the number of periods p, the elements of the characteristic matrix oscillate between the same high and low values. If, by proper dimensioning, we assure that the high and low values of the elements of the characteristic matrix (Eq. 3.3) result in high and acceptable transmittance values we can use as many periods as we want without substantially reducing transmittance. The region where this situation exists is called a *passband*.

2. N_{ABA} is *imaginary*. Now

$$M = \begin{bmatrix} \cosh p\Gamma_{ABA} & \dfrac{i \sinh p\Gamma_{ABA}}{N_{ABA}} \\ iN_{ABA} \sinh p\Gamma_{ABA} & \cosh p\Gamma_{ABA} \end{bmatrix}$$

If we increase the number of periods p, the matrix elements no longer oscillate but continuously increase in value. According to Eq. 2.40, the increase of all the matrix elements leads to a decrease in transmittance. We now have a situation where the transmittance continuously decreases whenever we add another period. The region where this situation exists is called a *stopband*.

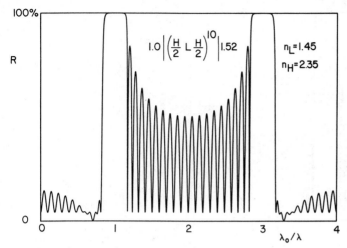

Figure 3.3. Reflectance of the design $1.0 \mid (\text{H/2 L H/2})^{10} \mid 1.52$ with $n_H = 2.35$ and $n_L = 1.45$.

Figures 3.3 and 3.4 show the transmittance of a multilayer in which the periods of Figs. 3.1 and 3.2 were repeated 10 times.

3.1.2. Invariances of equivalent layers†

We can use the relations derived in Sec. 2.8 to establish the following invariances of the equivalent indices and thicknesses:

1. *Multiply all indices with a constant factor.*   Comparing the definitions (Eqs. 3.1 and 3.2) with Eq. 2.44 we can conclude

$$N(n) = cN\left(\frac{n}{c}\right) \tag{3.4}$$

$$\Gamma(n) = \Gamma\left(\frac{n}{c}\right) \tag{3.5}$$

or, in words, the *equivalent index* of a symmetric combination of layers is c times the *equivalent index* of the same combination but with all indices replaced by n/c. The equivalent thicknesses are equal.

Example $N(\text{Ge-ZnS}) = 1.75N\,(\text{TiO}_2\text{-CaF}_2)$ and $\Gamma(\text{Ge-ZnS}) = \Gamma(\text{TiO}_2\text{-CaF}_2)$, assuming that $n_{\text{Ge}} = 4.0$, $n_{\text{ZnS}} = 2.2$, $n_{\text{TiO}_2} = 2.3$, and $N_{\text{CaF}_2} = 1.26$.

---

†From Thelen.[2a]

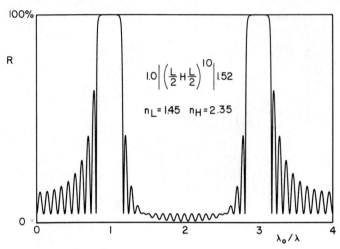

Figure 3.4. Reflectance of the design $1.0 \mid (L/2 \, H \, L/2)^{10} \mid 1.52$ with $n_H = 2.35$ and $n_L = 1.45$.

2. *Replace all indices by reciprocal values.* Let us now compare the definitions (Eqs. 3.1 and 3.2) with Eq. 2.46. We find correspondingly

$$N(n) = N\!\left(\frac{1}{n}\right) \tag{3.6}$$

$$\Gamma(n) = \Gamma\!\left(\frac{1}{n}\right) \tag{3.7}$$

or, in words, the *equivalent index and thickness* of a symmetric combination of layers *do not change* when the *indices of refraction* of all the layers are replaced by the reciprocal values.

Example   Applying Eqs. 3.4 to 3.7 in combination, we find that

$$N(n_H, n_L) = \frac{n_H n_L}{N(n_L, n_H)} \quad \text{and} \quad \Gamma(n_H, n_L) = \Gamma(n_L, n_H) \tag{3.8}$$

no matter how complex the symmetric two-material equivalent structure is.

3.1.3. An approximation for the equivalent thickness†

Equation 2.38 gives an expression for the complex amplitude transmittance $\vec{T}$ of a multilayer. Let us apply this equation to an equivalent

---

†From Epstein.[3]

layer. Setting $\vec{T}_e = |\vec{T}_e| \, e^{-i\theta}$, assuming equal massive media on both sides of the symmetric multilayer ($n_0 = n_S$), taking reciprocals of both sides, and rearranging slightly, we obtain

$$\frac{M_{11} + M_{22}}{2} + \frac{i(n_0 M_{12} + M_{21}/n_0)}{2} = \frac{e^{i\theta}}{|\vec{T}_e|}$$

or, equating the real parts,

$$\frac{M_{11} + M_{22}}{2} = \frac{\cos\theta}{|\vec{T}_e|} \tag{3.9}$$

Finally, with Eq. 3.2, we can write

$$\cos\Gamma = \frac{\cos\theta}{|\vec{T}_e|} \tag{3.10}$$

For simple equivalent layers where the directly transmitted light is the most important and zigzag reflections may be neglected, we can set $|\theta|$ equal to the sum of all the phase thicknesses ϕ of the multilayer, $|\theta| = \Sigma\phi$, and with Eq. 3.10

$$\Gamma \approx \Sigma\phi$$

A more rigorous proof of these relationships was given in an unpublished book by Epstein[4].

3.1.4. Single-layer synthesis
with equivalent layers

In Fig. 3.5 we show the equivalent index of the three-layer combination aH 2bL aH with $n_H = 2.35$, $n_L = 1.45$, and varying shift factors a and b. In order to center the coatings at $\lambda_0/\lambda = 1$, we set a + b = 1. We can see that it is possible to synthetically generate quarter-wave layers with any index of refraction between n_L and n_H. For thicker layers, the synthetic generation of indices outside the range set by n_L and n_H is even possible.

For the three-layer combination aA 2bB aA, expressions are available to compute the shift factors a and b for given n_A, n_B, and N_x and Γ_x of the film to be synthesized (Miller,[5] Ohmer[6]).

Multiplication of the characteristic matrices (Eq. 2.17) of the three layers aA 2bB aA leads to

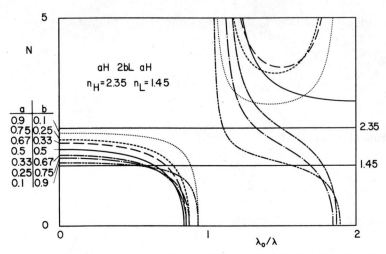

Figure 3.5. The equivalent index of the combination aH 2bL aH with $n_H = 2.35$ and $n_L = 1.45$ (a = 0.9, 0.75, 0.67, 0.5, 0.33, 0.25, 0.1 and b = 1 − a).

$$M_{11} = M_{22}$$

$$= \cos 2\phi_A \cos 2\phi_B - \tfrac{1}{2}[(n_A/n_B + n_B/n_A) \sin 2\phi_A \sin 2\phi_B] \quad (3.11)$$

$$n_A M_{12} = \sin 2\phi_A \cos 2\phi_B + \tfrac{1}{2}[(n_A/n_B + n_B/n_A) \cos 2\phi_A \sin 2\phi_B$$

$$+ (n_A/n_B - n_B/n_A) \sin 2\phi_B)] \quad (3.12)$$

$$\frac{M_{21}}{n_A} = \sin 2\phi_A \cos 2\phi_B + \tfrac{1}{2}[(n_A/n_B + n_B/n_A) \cos 2\phi_A \sin 2\phi_B$$

$$- (n_A/n_B - n_B/n_A) \sin 2\phi_B \quad (3.13)$$

Subtracting Eq. 3.13 from Eq. 3.12, we obtain

$$n_A M_{12} - \frac{M_{21}}{n_A} = \left(\frac{n_A}{n_B} - \frac{n_B}{n_A}\right) \sin 2\phi_B \quad (3.14)$$

It follows from Eqs. 3.1 and 3.2 that $M_{12} = (\sin \Gamma_x)/N_x$ and $M_{21} = N_x \sin \Gamma_x$. Inserting these relations into Eq. 3.14 yields

$$\sin 2\phi_B = \frac{\sin \Gamma_x (n_A/N_x - N_x/n_A)}{(n_A/n_B - n_B/n_A)} \quad (3.15)$$

TABLE 3.1   a and b for N(aA 2bB aA) with
$n_A = 2.35$, $n_B = 1.45$

N	a	b
2.35	0.500	0
2.25	0.469	0.028
2.15	0.437	0.057
2.05	0.402	0.088
1.95	0.365	0.122
1.85	0.325	0.160
1.75	0.279	0.203
1.65	0.227	0.256
1.55	0.159	0.325
1.45	0	0.500

With Eq. 3.2, Eq. 3.11 is of the type

$$w = u \sin 2\phi_A + v \cos 2\phi_A$$

where
$$u = -\left(\frac{n_A/n_B + n_B/n_A}{2}\right)\sin 2\phi_B$$

and
$$v = \cos 2\phi_B \qquad w = \cos \Gamma_x$$

We can solve this equation for ϕ_A:

$$\sin 2\phi_A = \frac{uw}{u^2 + w^2} \pm \sqrt{\frac{uv}{(u^2+w^2)^2} - \frac{w^2-v^2}{(u^2+v^2)^2}} \qquad (3.16)$$

Equations 3.15 and 3.16 are very tricky because it is difficult to establish the right quadrant for the inverse trigonometric functions. Equation 3.10 can be of great help in this task (Dobrowolski[7]). Table 3.1 gives a set of solutions for quarter-wave layers synthesized with $n_A = 2.35$ and $n_B = 1.45$.

Examples The single film $N_x = 1.95$ and $QWOT_x = 500$ nm is equivalent to the combination $n_A = 2.35$, $QWOT_A = 0.365 \times 500$ nm $= 182.5$ nm; $n_B = 1.45$, $QWOT_B = 2 \times 0.122 \times 500$ nm $= 122$ nm; $n_A = 2.35$, $QWOT_A = 0.365 \times 500$nm $= 182.5$ nm. Also, since $N_x = 1.95 = 1.45 \times 2.35/1.75$, it is, with Eq. 3.8, also equivalent to the combination (Table 3.1, a and b for $N = 1.75$) $n_A = 1.45$, $QWOT_A = 0.279 \times 500$ nm $= 139.5$ nm; $n_B = 2.35$, $QWOT_B = 2 \times 0.203 \times 500$ nm $= 203.5$ nm; $n_A = 1.45$; $QWOT_A = 0.279 \times 500$ nm $= 139$ nm.

3.1.5. Equivalent layers for half/quarter periodic multilayers

Figure 3.6 gives the equivalent index of the four possible ways to construct the base period of a periodic multilayer with alternating half-

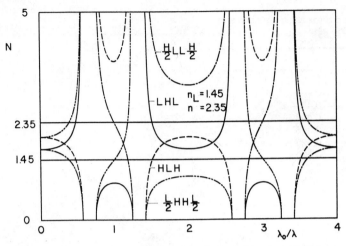

Figure 3.6. The equivalent index N for the four combinations H/2 LL H/2, L/2 HH L/2, LHL, HLH, with $n_L = 1.45$ and $n_H = 2.35$.

wave- and quarter-wave-thick layers. Equivalent indices of additional, more complex combinations are given later.

3.1.6. Image equivalent layers
of nonsymmetric sequences†

Let us assume a nonsymmetric sequence of layers, aA bB cC, with characteristic matrix $M_{11}, M_{12}, M_{21}, M_{22}$. If we supplement this structure with its images to the left or to the right we generate the following two new structures:

With left image: cC bB aA aA bB cC

and

With right image: aA bB cC cC bB aA

With Eqs. 2.27, 2.28, and 3.1 we can determine the equivalent indices of the imaged structures to be

$$N_{+\text{left image}} = \sqrt{\frac{M_{11}M_{21}}{M_{22}M_{12}}} \qquad (3.17)$$

and
$$N_{+\text{right image}} = \sqrt{\frac{M_{22}M_{21}}{M_{11}M_{12}}} \qquad (3.18)$$

---

†From Epstein.[8]

Let us now imagine that the original sequence of layers aA bB cC is embedded between two massive media with indices equal to the equivalent indices of the structures (Eq. 3.17 and Eq. 3.18)

$$n_0 = N_{+\text{right image}} \quad \text{and} \quad n_S = N_{+\text{left image}}$$

We find that, after inserting these values into Eq. 2.40,

$$T = \frac{4}{2 + M_{11}M_{22} + M_{11}M_{22} + M_{12}M_{21} + M_{12}M_{21}} = \frac{4}{2 + 2}$$
$$= 1 = 1 - R$$

or $R = 0$

We conclude that the *reflectance* of a nonsymmetric sequence of layers embedded between two massive media is zero as long as the indices of the massive media equal the equivalent indices of the sequence supplemented by its respective image.

Example $R \equiv 0$ for $N_{\text{LHHL}} \mid \text{H L} \mid N_{\text{HLLH}}$.

3.1.7. Equivalent layers for periodic nonsymmetric sequences

In Sec. 3.1 we established that a sequence of layers is equivalent to a single film whenever it is symmetric. When the sequence is nonsymmetric it is equivalent to two films (Herpin[1]). Let us establish this two-layer equivalence. M_{11}, M_{12}, M_{21}, and M_{22} are the elements of the characteristic matrix of the nonsymmetric sequence ($M_{11} \neq M_{22}$). We equate this matrix with the characteristic matrix of a two-layer film with indices N_1, N_2 and thicknesses Γ_1, Γ_2.

$$\begin{bmatrix} M_{11} & iM_{12} \\ iM_{21} & M_{22} \end{bmatrix}$$

$$= \begin{bmatrix} \cos \Gamma_1 \cos \Gamma_2 - \dfrac{N_2 \sin \Gamma_1 \sin \Gamma_2}{N_1} & i\left(\dfrac{\cos \Gamma_1 \sin \Gamma_2}{N_2} + \dfrac{\sin \Gamma_1 \cos \Gamma_2}{N_1} \right) \\[4mm] i(N_1 \sin \Gamma_1 \cos \Gamma_2 + N_2 \cos \Gamma_1 \sin \Gamma_2) & \cos \Gamma_1 \cos \Gamma_2 - \dfrac{N_1 \sin \Gamma_1 \sin \Gamma_2}{N_2} \end{bmatrix}$$

$$(3.19)$$

By equating the matrix elements we obtain

$$N_2^2 = \frac{N_1^2 (M_{11}^2 - 1) + M_{21}^2}{1 - M_{22}^2 - N_1^2 M_{12}^2}$$

$$\cos (\Gamma_1 + \Gamma_2) = \frac{N_1 M_{11} + N_2 M_{22}}{N_1 + N_2} \qquad (3.20)$$

$$\cos (\Gamma_1 - \Gamma_2) = \frac{N_1 M_{11} - N_2 M_{22}}{N_1 - N_2}$$

One of the parameters of the two-layer equivalent film can be selected at will. It most conveniently is N_1. By equating it with one of the outer-

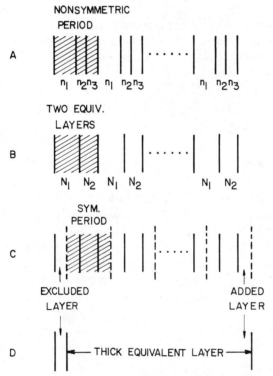

Figure 3.7. Process of finding an equivalent layer for a periodic multilayer with nonsymmetric periods: (A) Periodic multilayer with the nonsymmetric period. (B) Replacing the period with an equivalent two-layer system. (C) Exclusion of one-half of the first equivalent layer and addition of one-half of the first equivalent layer to the right side. (D) Finding a single equivalent layer for the newly formed symmetric period. The excluded layer has to be incorporated into the matching network.

TABLE 3.2   Equivalent Index N_2, Equivalent Thicknesses Γ_1, Γ_2, and $M_{11} + M_{22}$ of the Four-Layer Nonsymmetric Period | 0.6445A 0.6445B 0.3908A 0.3202B| with n_A = 1.45, n_B = 2.35, and N_1 = 1.45

λ_0/λ	N_2	Γ_1	Γ_2	$M_{11} + M_{22}$
1.000	—	—	—	−2.036
1.050	—	—	—	−1.978
1.100	1.007	1.750	0.286	−1.867
1.150	1.157	1.546	0.505	−1.704
1.250	1.156	1.226	0.844	−1.242
1.300	1.090	1.121	0.959	−0.952
1.350	0.986	1.050	1.044	−0.632
1.400	0.828	1.127	0.987	−0.290
1.450	0.566	1.207	0.944	−0.067
1.500				0.431

layer indices of the nonsymmetric sequence or with one of the massive media indices, matching might be simplified. It can also be used to prevent N_2 to become imaginary in a passband.

At first sight, the two-layer equivalence appears to be of little help in matching a multilayer with nonsymmetric periods to the adjacent massive media. With symmetric periods it was possible to treat the multilayer as a thick single film while two-layer equivalent layers keep their alternating index character. But by adding a layer and by

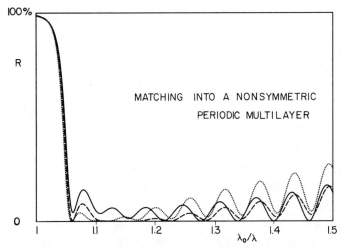

Figure 3.8. Reflectance of the periodic multilayer with nonsymmetric period: 1 | (0.6445A 0.6445B 0.3908A 0.3202B)^{15} | 1.52 (solid curve), 1 | 0.0964A (0.6445A-0.6445B 0.3908A 0.3202B)^{15} 0.773A 0.87D | 1.52 (dashed curve), and 1 | 0.0339A (0.6445A 0.6445B 0.3908A 0.3202B)^{15} 0.8751A 0.9091D | 1.52 (dotted curve) with n_A = 1.45, n_B = 2.35, n_D = 1.758, and n_{D2} = 1.882.

splitting off another we can also find for the nonsymmetric sequence a single-film equivalent layer—the equivalent of the equivalent so to speak. Figure 3.7 on p. 52 shows the procedure graphically.

As an example, let us try to improve the transmittance in the pass-band of a nonpolarizing edge filter (Fig. 9.21). The nonsymmetric period consists of the four layers | 0.6445A 0.6445B 0.3908A 0.3202B | with $n_A = 1.45$ and $n_B = 2.35$. Table 3.2 gives the equivalent index N_2, the equivalent thicknesses Γ_1, Γ_2, and $M_{11} + M_{22}$ of the equivalent two-layer structure. N_1 was made equal to the index of the first layer of the original period. Equations 3.20 were used.

We select as match point $\lambda_0/\lambda = 1.15$. From Table 3.2 we determine the following two-layer equivalent structure: | 1.546A 0.505C | with $n_A = 1.45$ and $n_C = 1.157$. By excluding the first half of the first equivalent layer and adding it to the end we arrive at the symmetric structure | 0.773A 0.505C 0.773A | with equivalent index $N = 2.033$ at

TABLE 3.3 Properties of the Chebyshev Polynomials $T_m(X)$

Definition:
$$T_m(x) = \cos(m \arccos x)$$
$$= 2^{m-1} + \left(2\begin{bmatrix} m - 1 \\ 1 \end{bmatrix} - \begin{bmatrix} m - 2 \\ 1 \end{bmatrix}\right) 2^{m-3} x^{m-2}$$
$$+ \left(2\begin{bmatrix} m - 2 \\ 2 \end{bmatrix} - \begin{bmatrix} m - 3 \\ 2 \end{bmatrix}\right) 2^{m-5} x^{m-4} \cdots$$

$T_0(x) = 1$

$T_1(x) = x$

$T_2(x) = 2x^2 - 1$

$T_3(x) = 4x^3 - 3x$

$T_4(x) = 8x^4 - 8x^2 + 1$

$T_5(x) = 16x^5 - 20x^3 + 5x$

$T_6(x) = 32x^6 - 48x^4 + 18x^2 - 1$

$T_7(x) = 64x^7 - 112x^5 + 56x^3 - 7x$

$T_8(x) = 128x^8 - 256x^6 + 160x^4 - 32x^2 + 1$

$T_9(x) = 256x^9 - 576x^7 + 432x^5 - 120x^3 + 9x$

$T_{10}(x) = 512x^{10} - 1280x^8 + 1120x^6 - 400x^4 + 50x^2 + 1$

$T_{11}(x) = 1024x^{11} - 2816x^9 + 1816x^7 - 1232x^5 + 220x^3 - 11x$

$T_{12}(x) = 2048x^{12} - 6144x^{10} + 6912x^8 - 3584x^6 + 840x^4 - 72x^2 + 1$

Recurrence relation $T_{m+1}(x) - 2xT_m(x) + T_{m-1}(x) = 0$

Multiplication formula $T_{m1}(x)T_{m2}(x) = \dfrac{T_{m1+m2}(x) + T_{m1-m2}(x)}{2}$

Functional equation $T_{m1}(T_{m2}(x)) = T_{m2}(T_{m1}(x)) = T_{m1 \times m2}(x)$

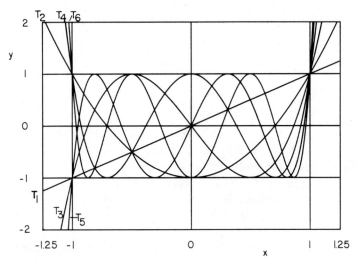

Figure 3.9. Plot of the Chebyshev polynomials $T_m(x)$ for $m = 1$ to 6 and $-1.25 < x < 1.25$

$\lambda_0/\lambda = 1.15$ (Prob. 3.1). The excluded layer $\mid 0.773A \mid$ provides a good match to air if we add the layer $\mid 0.0964A \mid$ to it which makes it a quarter wave at $\lambda_0/\lambda = 1.15$. A layer with $n_D = \sqrt{1.52 \times 2.003} = 1.745$ provides a good match to the substrate. In Fig. 3.8 we show the result of our matching process. The dotted curve shows the result of the same procedure carried out at $\lambda_0/\lambda = 1.1$.

Obviously, this is not an easy process. Most of the time, the excluded layer does not fit in as nicely. Sometimes, elimination of imaginary equivalent indices through variation of N_1 is very difficult.

3.2. Synthesis with Chebyshev Polynomials

3.2.1 Equal ripple prototype antireflection coatings

When all the m layers of a multilayer have equal phase thickness ϕ the characteristic matrix (Eq. 2.22) will have the form

$$\mathbf{M} = \begin{pmatrix} a_0 \cos^m\phi + a_2 \cos^{m-2}\phi + \cdots & i \sin\phi \, (a_1 \cos^{m-1}\phi + a_3 \cos^{m-3}\phi + \cdots) \\ i \sin\phi \, (b_1 \cos^{m-1}\phi + b_3 \cos^{m-3}\phi + \cdots) & b_0 \cos^m\phi + b_2 \cos^{m-2}\phi + \cdots \end{pmatrix} \quad (3.21)$$

where the coefficients $a_0, a_1, \ldots$ and $b_0, b_1, \ldots$ are only functions of the refractive indices of the individual layers.

If we insert the elements of the characteristic matrix (Eq. 3.21) into Eq. 2.40 we obtain for the reflectance:

$$R = 1 - T$$

$$= 1 - \frac{1}{c_0 \cos^{2m}\phi + c_1 \cos^{2m-1}\phi + c_3 \cos^{2m-2}\phi + \cdots + c_{2m}}$$

$$= \frac{P\ (\cos\phi)}{C + P\ (\cos\phi)} \tag{3.22}$$

where the coefficients c_0, c_1, . . . , c_{2m} are functions of the refractive indices of the films, the incident medium, or the substrate; C is a constant; and P stands for "polynomial of."

In polynomial synthesis, the polynomial of Eq. 3.22 is made equal to a polynomial whose characteristics fit a desired transmittance function. Equations between the coefficients and the refractive indices are established and solved (Pohlack,[9] Young,[10] Seeley[11]).

Due to their equal ripple character Chebyshev polynomials $T_m(x)$ are used [please note the nomenclature conflict between T = transmittance and $T_m(x)$ = Chebyshev polynomial of the first kind of order m]. Table 3.3 and Fig. 3.9 list the first 12, resp. 6, polynomials. For $-1 \leqslant x \leqslant +1$, $T_m(x)$ oscillates between the values $+1$ and -1 m times. Beyond $x = \pm 1$, $T_m(x)$ will increase or decrease steadily and rapidly towards $+\infty$.

Three parameters completely describe an equal ripple antireflection coating:

1. The passband ripple ρ as the maximum reflectance in the region of equal ripple performance
2. The fractional bandwidth $W = 2(\lambda_2 - \lambda_1)/(\lambda_2 + \lambda_1)$, where λ_1 is the minimum and λ_2 the maximum wavelength of equal ripple performance
3. The substrate index to incident medium index ratio n_S/n_0 (see Sec. 2.8)

In order to avoid negative transmittances we use $T_m^2(x)$ instead of $T_{2m}(x)$. This leads to

$$R = \frac{\rho T_m^2(x)}{1 + \rho T_m^2(x)} \tag{3.23}$$

For $-1 \leqslant x \leqslant +1$: $R_{max} = \rho/(1 + \rho) \approx \rho$ and $R_{min} = 0$.

In order to relate x to ϕ we set $x = \eta \cos\phi$. Let λ_1 and λ_2 be the wavelengths which correspond to $x = +1$ and $x = -1$ (defining the region of equal ripple performance). Then (Eqs. 2.5 and 2.56)

$$\frac{1}{\eta} = \cos\frac{2\pi n d}{\lambda_1} = \cos\left(\frac{\pi}{2}\frac{\lambda_0}{\lambda_1}\right)$$

Since $\phi \sim 1/\lambda$, λ_1, λ_2, and λ_0 are connected through the relationship

$$\frac{2}{\lambda_0} = \frac{1}{\lambda_1} + \frac{1}{\lambda_2}$$

We can eliminate λ_0 and obtain

$$\frac{1}{\eta} = \cos \frac{\pi \lambda_2/\lambda_1}{1 + \lambda_2/\lambda_1}$$

$$= \sin \frac{\pi W}{4}$$

Equation 3.23 now becomes

$$R = \frac{\rho T_m^2 \, [\cos \phi/\sin (W\pi/4)]}{1 + \rho T_m^2 \, [\cos \phi/\sin (W\pi/4)]} \tag{3.24}$$

The third parameter, n_S/n_0 (substrate index to incident medium index ratio), is no longer free if the other two are fixed. At $\lambda \to \infty$ ($\cos \phi = 1$), the reflectance of the multilayer should be equal to the reflectance of the bare substrate. Consequently

$$\frac{(n_s - n_0)^2}{(n_s + n_0)^2} = \frac{\rho T_m^2 \, [1/\sin (W\pi/4)]}{1 + \rho T_m^2 \, [1/\sin (W\pi/4)]}$$

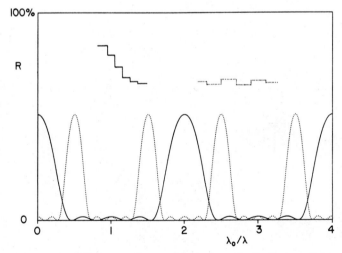

Figure 3.10. Reflectance of an equal ripple antireflection coating (ERAR) (solid curve), with $n_S/n_0 = 4$ and $\lambda_2/\lambda_1 = 3$. Equations 3.24 and 3.25 were used for the computation. Also, the reflectance of the corresponding equal ripple filter (ERF) (dotted curve) (Sec. 3.2.2) is shown. The diagrams in the upper part of the figure show the respective refractive indices as a function of the optical thicknesses.

or
$$\rho = \frac{(n_s/n_1 - 1)^2}{4n_s/n_0} \frac{1}{T_m^2 \, [1/\sin \, (W\pi/4)]}$$
(3.25)

The general shape of $R = f(\lambda_0/\lambda)$ (Eq. 3.24) is given in Fig. 3.10 (solid curve).

TABLE 3.4 Collin's Formulas for the Refractive Indices of Equal Ripple Antireflection Coatings

Two Layers

Maximally flat: $\dfrac{n_1}{n_0} = \sqrt{\dfrac{n_s}{n_0}}$

Equal ripple: $\dfrac{n_1}{n_0} = \left(\sqrt{C^2 + \dfrac{n_s}{n_0}} + C \right)^{1/2}$

where $C = \dfrac{n_s/n_0 - 1}{2(2\eta^2 - 1)}$

Both: $n_2 n_1 = n_0 n_s$ $\dfrac{1}{\eta} = \sin \dfrac{\pi W}{4}$

Three Layers

Maximally flat: $\left(\dfrac{n_1}{n_0} \right) + 2\sqrt{\dfrac{n_s}{n_0}} \left(\dfrac{n_1}{n_0} \right) - \dfrac{2\sqrt{n_s/n_0}}{n_1/n_0} - \dfrac{n_s n_0}{(n_1/n_0)^2} = 0$

Equal ripple: $\left(\dfrac{n_1}{n_0} \right) + 2\sqrt{\dfrac{n_s}{n_0}} \left(\dfrac{n_1}{n_0} \right) - \dfrac{2\sqrt{n_s/n_0}}{n_1/n_0} - \dfrac{n_s/n_0}{(n_1/n_0)^2} = \dfrac{3(n_s/n_0 - 1)}{4\eta^2 - 3}$

Both: $n_2 = \sqrt{n_0 n_s}$ $n_3 n_1 = n_0 n_s$

Four Layers

Maximally flat: $\dfrac{n_1}{n_0} = A_1 \left(\dfrac{n_s}{n_0} \right)^{1/8}$ $\dfrac{n_2}{n_1} = \left(\dfrac{n_s}{n_0} \right)^{1/4}$

$\dfrac{n_3}{n_2} = \dfrac{(n_s/n_0)^{1/4}}{A_1^2}$ $n_4 n_1 = n_0 n_s$

where $\dfrac{1}{A_1^2} - A_1^2 = 2\dfrac{(n_s/n_0)^{1/4} - 1}{(n_s/n_0)^{1/4} + 1}$

Equal ripple: $\dfrac{n_1}{n_0} = \sqrt{\left(\dfrac{n_s}{n_0} \right) \left[B + \left(B^2 + \dfrac{n_0 A^2}{n_s} \right)^{1/2} \right]}$

$\dfrac{n_2}{n_1} = \dfrac{1}{A}$

$n_3 n_2 = n_0 n_s$ $n_4 n_1 = n_0 n_s$

where $A^2 = (1 - n_0/n_s)/2t_1 t_2) + \sqrt{(1 - n_0/n_s)^2/(4t_1^2 t_2^2)} + n_0 n_s$

$2B = [A/(A + 1)]^2[(t_1 + t_2)(A^2 - n_0/A^2 n_s) - 2A$
$\qquad + 2n_0/An_s]$

$t_1 = 2\sqrt{2}\eta^2/(\sqrt{2} + 1) - 1$

$t_2 = 2\sqrt{2}\eta^2/(\sqrt{2} - 1) - 1$

TABLE 3.5 Refractive Indices n_1/n_0 of the First Layer of Two-Layer Equal Thickness Combinations†

n_s/n_0	λ_2/λ_1								
	1.000	1.500	1.750	2.000	2.500	3.000	4.000	6.000	10.00
1.100	1.024	1.025	1.026	1.028	1.030	1.032	1.036	1.041	1.045
1.200	1.047	1.049	1.051	1.053	1.058	1.063	1.070	1.080	1.088
1.300	1.068	1.071	1.074	1.078	1.085	1.091	1.102	1.117	1.129
1.400	1.088	1.092	1.096	1.101	1.110	1.119	1.133	1.152	1.169
1.500	1.107	1.112	1.117	1.123	1.134	1.145	1.163	1.186	1.207
1.600	1.125	1.131	1.137	1.144	1.157	1.170	1.191	1.219	1.244
1.700	1.142	1.150	1.156	1.164	1.179	1.194	1.219	1.251	1.279
1.800	1.158	1.167	1.175	1.183	1.201	1.217	1.245	1.282	1.314
1.900	1.174	1.184	1.192	1.202	1.221	1.240	1.271	1.311	1.347
2.000	1.189	1.200	1.209	1.220	1.241	1.261	1.295	1.340	1.380
2.200	1.218	1.230	1.241	1.254	1.279	1.303	1.343	1.396	1.442
2.400	1.245	1.259	1.271	1.285	1.314	1.342	1.388	1.449	1.502
2.600	1.270	1.286	1.300	1.316	1.348	1.379	1.431	1.500	1.560
2.800	1.294	1.311	1.327	1.344	1.380	1.414	1.472	1.548	1.615
3.000	1.316	1.335	1.352	1.371	1.411	1.448	1.512	1.595	1.668
3.500	1.368	1.391	1.412	1.435	1.482	1.528	1.605	1.706	1.794
4.000	1.414	1.441	1.465	1.492	1.547	1.600	1.691	1.809	1.911
4.500	1.456	1.487	1.514	1.545	1.608	1.668	1.771	1.905	2.021
5.000	1.495	1.529	1.560	1.594	1.664	1.732	1.847	1.997	2.126
6.000	1.565	1.606	1.642	1.683	1.768	1.850	1.988	2.167	2.319

†Between the massive media n_0 and n_S which reduce the reflectances to $R < R_{\max}$ from λ_1 to λ_2. The second layer has the index $n_2 = n_0 n_s/n_1$.

Explicit formulas for calculating the refractive indices for given n_S/n_0 and bandwidth W or λ_2/λ_1 were derived for two, three, and four layers by Collin.[12] In Table 3.4 we give adapted versions of Collin's formulas. Tabulated solutions to these formulas were published by Young.[13] These formulas are sometimes useful as a double check to the more general Riblet's procedure which will be discussed in Sec. 3.2.3.

Tables 3.5 to 3.8 give a set of values computed from Collin's formulas. Maximally flat (Butterworth) solutions as the limit of perfect match over vanishingly small bandwidth are included under $\lambda_2/\lambda_1 = 1$, $T_m^2(\eta \cos \phi)$ then becomes $(\eta \cos \phi)^{2m}$. Figure 3.11 on p. 63 gives the corresponding passband ripple ρ or maximum reflectances $R_{\max}$ computed from Eq. 3.25.

3.2.2. Equal ripple prototype filters

For large n_S/n_0 the antireflection coatings synthesized in the previous section make remarkably good bandpass filters. The dashed curve of Fig. 3.12 on p. 64 gives an example with $n_S/n_0 = 1000$, fifteen layers,

TABLE 3.6   Refractive Indices n_1/n_0 of the First Layer of Three-Layer Equal Thickness Combinations†

n_s/n_0	λ_2/λ_1								
	1.000	1.500	1.750	2.000	2.500	3.000	4.000	6.000	10.00
1.100	1.012	1.013	1.014	1.015	1.017	1.019	1.024	1.031	1.039
1.200	1.023	1.025	1.027	1.028	1.033	1.037	1.046	1.060	1.076
1.300	1.033	1.036	1.038	1.041	1.047	1.054	1.067	1.088	1.112
1.400	1.043	1.046	1.050	1.053	1.061	1.070	1.086	1.114	1.146
1.500	1.052	1.056	1.060	1.064	1.074	1.085	1.105	1.139	1.178
1.600	1.061	1.065	1.070	1.075	1.087	1.099	1.123	1.163	1.210
1.700	1.069	1.074	1.079	1.085	1.098	1.112	1.140	1.186	1.240
1.800	1.076	1.083	1.088	1.095	1.110	1.125	1.156	1.208	1.269
1.900	1.084	1.090	1.097	1.104	1.120	1.138	1.172	1.229	1.298
2.000	1.091	1.098	1.105	1.113	1.131	1.150	1.187	1.250	1.325
2.200	1.104	1.112	1.120	1.130	1.150	1.172	1.216	1.290	1.378
2.400	1.116	1.126	1.135	1.145	1.168	1.193	1.243	1.327	1.429
2.600	1.127	1.138	1.148	1.160	1.186	1.213	1.269	1.363	1.477
2.800	1.138	1.150	1.160	1.173	1.202	1.232	1.293	1.397	1.523
3.000	1.148	1.161	1.172	1.186	1.217	1.250	1.316	1.430	1.568
3.500	1.171	1.185	1.199	1.215	1.252	1.291	1.370	1.507	1.674
4.000	1.191	1.207	1.223	1.242	1.283	1.328	1.420	1.579	1.772
4.500	1.209	1.227	1.245	1.265	1.312	1.363	1.465	1.645	1.864
5.000	1.225	1.246	1.265	1.287	1.339	1.394	1.508	1.708	1.951
6.000	1.254	1.278	1.300	1.326	1.386	1.452	1.587	1.825	2.113

†Between the massive media n_0 and n_s which reduce the reflectance to $R < R_{max}$ from λ_1 to λ_2. The second layer has the index $n_2 = \sqrt{n_0 n_s}$ and the third layer $n_3 = n_0 n_s/n_1$.

and $W = (\lambda_2 - \lambda_1)/[(\lambda_2 + \lambda_1)/2] = 10\%$. Unfortunately, as filters, these designs would have to work between massive media with a huge difference in the refractive indices, because, for good filter performance, the ratio n_S/n_0 has to be very large. (n_S/n_0 is related to the transmittance in the stopband through the formula

$$\log T = \log 4 - \log \frac{n_S}{n_0} = 0.602 - \log \frac{n_S}{n_0} \qquad (3.26)$$

where the assumption is made that n_S/n_0 is large. Equation 3.26 is simply one minus the Fresnel reflectance. T is assumed as a fraction.) Young[10] proposed a set of index and thickness transformations to generate the same transmittance characteristic between equal (odd number of layers) or nearly equal (even number of layers) indices of the massive media surrounding the multilayer.

The characteristic matrix of the synthesized antireflection coatings, as described in the previous section, has the form

$$\mathbf{M} = \begin{bmatrix} \cos\phi & \dfrac{i\sin\phi}{n_1} \\ in_1\sin\phi & \cos\phi \end{bmatrix} \begin{bmatrix} \cos\phi & \dfrac{i\sin\phi}{n_2} \\ n_2\sin\phi & \cos\phi \end{bmatrix} \dots \begin{bmatrix} \cos\phi & \dfrac{i\sin\phi}{n_m} \\ in_m\sin\phi & \cos\phi \end{bmatrix}$$

$$(3.27)$$

With Eqs. 2.41, 2.42, and 2.43 Eq. 3.27 can be modified into

$$\mathbf{M} = \begin{bmatrix} 1 & 0 \\ 0 & n_1 \end{bmatrix} \begin{bmatrix} \cos\phi & i\sin\phi \\ i\sin\phi & \cos\phi \end{bmatrix} \begin{bmatrix} 1 & 0 \\ 0 & \dfrac{n_2}{n_1} \end{bmatrix} \begin{bmatrix} \cos\phi & i\sin\phi \\ i\sin\phi & \cos\phi \end{bmatrix} \begin{bmatrix} 1 & 0 \\ 0 & \dfrac{n_3}{n_2} \end{bmatrix} \dots$$

$$\begin{bmatrix} \cos\phi & i\sin\phi \\ i\sin\phi & \cos\phi \end{bmatrix} \begin{bmatrix} 1 & 0 \\ 0 & \dfrac{1}{n_m} \end{bmatrix} \quad (3.28)$$

Since we are interested in modifying the relationship of the multilayer relative to its optical environment, let us switch from characteristic to transfer matrices. With transformation Eq. 2.33 we have

$$\mathbf{Q} = 0.5 \begin{bmatrix} 1 & \dfrac{n_1}{n_0} \\ 1 & \dfrac{-n_1}{n_0} \end{bmatrix} \begin{bmatrix} \cos\phi & i\sin\phi \\ i\sin\phi & \cos\phi \end{bmatrix} \begin{bmatrix} 1 & 0 \\ 0 & \dfrac{n_2}{n_1} \end{bmatrix} \begin{bmatrix} \cos\phi & i\sin\phi \\ i\sin\phi & \cos\phi \end{bmatrix} \begin{bmatrix} 1 & 0 \\ 0 & \dfrac{n_3}{n_2} \end{bmatrix} \dots$$

$$\begin{bmatrix} \cos\phi & i\sin\phi \\ i\sin\phi & \cos\phi \end{bmatrix} \begin{bmatrix} 1 & 1 \\ \dfrac{n}{n_m} & \dfrac{-n_s}{n} \end{bmatrix} \quad (3.29)$$

Consider the thickness transformation $\phi \lozenge 90° + \phi$. The matrices containing ϕ in Eq. 3.29 would be transformed into

$$\begin{bmatrix} \cos\phi & i\sin\phi \\ i\sin\phi & \cos\phi \end{bmatrix} \Rightarrow \begin{bmatrix} -\sin\phi & i\cos\phi \\ i\cos\phi & -\sin\phi \end{bmatrix} \quad (3.30)$$

On the other hand, the two matrices of transformation Eq. 3.30 can be related to each other by the two identities

$$\begin{bmatrix} \cos\phi & i\sin\phi \\ i\sin\phi & \cos\phi \end{bmatrix} = \begin{bmatrix} 0 & -i \\ -i & 0 \end{bmatrix} \begin{bmatrix} -\sin\phi & i\cos\phi \\ i\cos\phi & -\sin\phi \end{bmatrix} \quad (3.31)$$

$$\begin{bmatrix} \cos\phi & i\sin\phi \\ i\sin\phi & \cos\phi \end{bmatrix} = \begin{bmatrix} -\sin\phi & i\cos\phi \\ i\cos\phi & -\sin\phi \end{bmatrix} \begin{bmatrix} 0 & -i \\ -i & 0 \end{bmatrix} \quad (3.32)$$

TABLE 3.7 Refractive Indices n_1/n_0 of the First Layer of Four-Layer Equal Thickness Combinations†

	$\lambda_2/\lambda_1 =$								
n_s/n_0	1.000	1.500	1.750	2.000	2.500	3.000	4.000	6.000	10.00
1.100	1.006	1.007	1.007	1.008	1.009	1.011	1.015	1.022	1.033
1.200	1.011	1.013	1.014	1.015	1.018	1.022	1.029	1.043	1.063
1.300	1.017	1.018	1.020	1.022	1.026	1.031	1.042	1.063	1.093
1.400	1.021	1.024	1.026	1.028	1.034	1.040	1.054	1.081	1.120
1.500	1.026	1.028	1.031	1.034	1.041	1.049	1.066	1.099	1.147
1.600	1.030	1.033	1.036	1.040	1.048	1.057	1.077	1.115	1.173
1.700	1.034	1.037	1.041	1.045	1.054	1.065	1.087	1.131	1.197
1.800	1.038	1.042	1.045	1.050	1.060	1.072	1.097	1.147	1.221
1.900	1.041	1.045	1.050	1.055	1.066	1.079	1.107	1.161	1.244
2.000	1.044	1.049	1.054	1.059	1.072	1.086	1.116	1.175	1.266
2.200	1.051	1.056	1.061	1.068	1.082	1.098	1.133	1.203	1.309
2.400	1.057	1.063	1.069	1.076	1.092	1.110	1.149	1.228	1.349
2.600	1.062	1.069	1.075	1.083	1.101	1.121	1.164	1.252	1.388
2.800	1.067	1.074	1.081	1.090	1.109	1.131	1.179	1.275	1.424
3.000	1.072	1.080	1.087	1.096	1.117	1.141	1.192	1.297	1.460
3.500	1.082	1.092	1.100	1.111	1.135	1.163	1.224	1.348	1.544
4.000	1.092	1.102	1.112	1.124	1.151	1.183	1.252	1.395	1.621
4.500	1.100	1.112	1.123	1.135	1.166	1.201	1.278	1.438	1.694
5.000	1.108	1.120	1.132	1.146	1.179	1.217	1.302	1.479	1.763
6.000	1.122	1.135	1.149	1.165	1.203	1.247	1.346	1.554	1.891

†Between the massive media n_0 and n_s which reduce the reflectance to $R < R_{max}$ from λ_1 to λ_2. The fourth layer has the index $n_4 = n_0 n_s/n_1$.

If we now make alternating substitutions into Eq. 3.29: Eq. 3.31 for the first ϕ-matrix, Eq. 3.32 for the second, Eq. 3.31 for the third, and so on, and observe that

$$\begin{bmatrix} 0 & -i \\ -i & 0 \end{bmatrix} \begin{bmatrix} 1 & 0 \\ 0 & \dfrac{n_k}{n_{k-1}} \end{bmatrix} \begin{bmatrix} 0 & -i \\ -i & 0 \end{bmatrix} = \dfrac{n_k}{n_{k-1}} \begin{bmatrix} 1 & 0 \\ 0 & \dfrac{n_{k-1}}{n_k} \end{bmatrix}$$

we obtain finally

$$\mathbf{Q} = C \begin{bmatrix} 1 & \dfrac{n_1}{n_0} \\ 1 & \dfrac{-n_1}{n_0} \end{bmatrix} \begin{bmatrix} -\sin\phi & i\cos\phi \\ i\cos\phi & -\sin\phi \end{bmatrix} \begin{bmatrix} 1 & 0 \\ 0 & \dfrac{n_1}{n_2} \end{bmatrix} \begin{bmatrix} -\sin\phi & i\cos\phi \\ i\cos\phi & -\sin\phi \end{bmatrix} \begin{bmatrix} 1 & 0 \\ 1 & \dfrac{n_3}{n_2} \end{bmatrix}$$

$$\begin{bmatrix} -\sin\phi & i\cos\phi \\ i\cos\phi & -\sin\phi \end{bmatrix} \begin{bmatrix} 1 & 0 \\ 0 & \dfrac{n_3}{n} \end{bmatrix} \cdots \begin{bmatrix} -\sin\phi & i\cos\phi \\ i\cos\phi & -\sin\phi \end{bmatrix} \mathbf{L} \quad (3.33)$$

TABLE 3.8   Refractive Indices n_2/n_0 of the Second Layer of Four-Layer Equal Thickness Combinations†

	$\lambda_2/\lambda_1 =$								
n_s/n_0	1.000	1.500	1.750	2.000	2.500	3.000	4.000	6.000	10.00
1.100	1.030	1.031	1.031	1.032	1.033	1.034	1.036	1.039	1.043
1.200	1.059	1.060	1.061	1.062	1.064	1.066	1.070	1.077	1.084
1.300	1.085	1.087	1.089	1.090	1.094	1.097	1.103	1.112	1.124
1.400	1.111	1.113	1.115	1.117	1.122	1.126	1.134	1.146	1.161
1.500	1.135	1.138	1.140	1.143	1.149	1.154	1.163	1.179	1.198
1.600	1.158	1.162	1.165	1.168	1.174	1.181	1.192	1.210	1.233
1.700	1.180	1.184	1.188	1.191	1.199	1.206	1.219	1.240	1.266
1.800	1.202	1.206	1.210	1.214	1.223	1.231	1.246	1.269	1.299
1.900	1.222	1.227	1.231	1.236	1.246	1.255	1.271	1.297	1.331
2.000	1.242	1.247	1.252	1.257	1.268	1.278	1.296	1.325	1.362
2.200	1.280	1.286	1.292	1.298	1.310	1.322	1.343	1.377	1.421
2.400	1.315	1.322	1.329	1.336	1.350	1.363	1.388	1.427	1.478
2.600	1.349	1.357	1.364	1.371	1.387	1.402	1.430	1.475	1.532
2.800	1.380	1.389	1.397	1.406	1.423	1.440	1.471	1.520	1.584
3.000	1.411	1.420	1.429	1.438	1.457	1.476	1.509	1.564	1.634
3.500	1.481	1.492	1.503	1.514	1.537	1.559	1.600	1.667	1.752
4.000	1.544	1.558	1.570	1.583	1.610	1.636	1.684	1.762	1.862
4.500	1.603	1.618	1.632	1.647	1.677	1.707	1.761	1.850	1.964
5.000	1.657	1.674	1.689	1.706	1.740	1.773	1.834	1.934	2.062
6.000	1.755	1.776	1.794	1.814	1.854	1.894	1.967	2.088	2.242

†Between the massive media n_0 and n_s which reduce the reflectance to $R < R_{max}$ from λ_1 to λ_2. The third layer has the index $n_3 = n_0 n_s/n_2$.

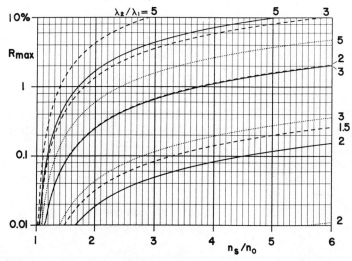

Figure 3.11.   Maximum reflectance R_{max} of the designs given in Tables 3.5 to 3.7: two layers (dashed curves), three layers (solid curves), and four layers (dotted curves). (λ_2/λ_1 is a parameter.)

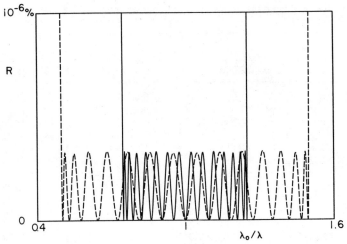

Figure 3.12. Reflectance of the equal ripple antireflection coating (dashed curve) and the equal ripple filter (solid curve) of Table 3.9.

The last matrix is different for odd or even numbers of layers. For an odd number of layers it is

$$\mathbf{L} = \begin{bmatrix} 1 & 1 \\ \dfrac{n_s}{n_m} & -\dfrac{n_s}{n_m} \end{bmatrix} \tag{3.34}$$

and for an even number of layers it is

$$\mathbf{L} = \begin{bmatrix} 1 & -i \\ -i & 1 \end{bmatrix} \begin{bmatrix} 1 & 1 \\ \dfrac{n_s}{n_m} & -\dfrac{n_s}{n_m} \end{bmatrix} = C_1 \begin{bmatrix} 1 & -1 \\ \dfrac{n_m}{n_s} & \dfrac{n_m}{n_s} \end{bmatrix} \tag{3.35}$$

The factors C and C_1 cancel out when the values for the elements of matrix $\mathbf{Q}$ (Eq. 3.33) are inserted into the reflectance formula (Eq. 2.35). Also, the fact that the matrix on the right side of Eq. 3.34 and the last matrix of Eq. 3.35 differ by a minus sign in the last column is of no consequence for the determination of the reflectance since $E^-(z_{m+1} + \delta) = 0$.

Let us compare Eqs. 3.29 and 3.33. In both equations the indices of refraction appear in ratios only. Yet in Eq. 3.29 they progress as

$$\frac{n_1}{n_0}, \quad \frac{n_2}{n_1}, \quad \frac{n_3}{n_2}, \quad \frac{n_4}{n_3}, \quad \cdots, \quad \frac{n_s}{n_m} \tag{3.36}$$

while in Eq. 3.33 they progress as

$$\frac{n_1}{n_0}, \quad \frac{n_1}{n_2}, \quad \frac{n_3}{n_2}, \quad \frac{n_3}{n_4}, \quad \ldots \tag{3.37}$$

The last index ratio is n_S/n_m for m odd and n_m/n_S for m even. We observe that in the progression (3.37) every second index ratio is reversed. Young[14] called this new filter the "half-wave filter." In thin film optics this name is somewhat misleading. We prefer to use the term *equal ripple antireflection coatings (ERAR)* for the prototypes introduced in Sec. 3.2.1 and *equal ripple filter (ERF)* for the prototypes introduced in this section. See also Fig. 3.10.

Equation 3.37 describes the new filter in terms of index ratios. In absolute terms the indices progress in the following way:

$$n_0, \quad n_1, \quad \frac{n_1^2}{n_2}, \quad \frac{n_1^2 n_3}{n_2^2}, \quad \frac{n_1^2 n_3^2}{n_2^2 n_4}, \quad \frac{n_1^2 n_3^2 n_5}{n_2^2 n_4^2}, \quad \ldots \tag{3.38}$$

For an odd number of layers, the indices of refraction of the massive entrance and exit media are equal: $n_S = n_0$. This follows from Eq. 3.37 and $n_k n_{m-k} = n_S/n_0$ (which is valid for equal ripple antireflection coatings but no longer valid for equal ripple filters). For an even number of layers n_0 and n_S are different. n_s must be calculated using the scheme of Eq. 3.37.

In Table 3.9 we compare the indices of refraction of the equal ripple antireflection coating (dashed curve) of Fig. 3.12 with the equal ripple filter (solid curve) derived from it. The fact that the bandwidth of the equal ripple filter is only one-half the bandwidth of the equal ripple antireflection coating is the result of the thickness transformation $\phi \lozenge 90° + \phi$.

TABLE 3.9   Indices of Refraction of a 15-Layer Equal Ripple Antireflection Coating (I) and an Equal Ripple Filter (II) Derived from It (Eq. 3.33)

Layer number	0	1	2	3	4	5	6	7
Indices (I)	1.000	1.006	1.043	1.181	1.573	2.595	5.267	12.42
Index ratios (I)	1.000	1.006	1.037	1.132	1.332	1.650	2.029	2.359
Indices (II)	1.000	1.006	0.970	1.098	0.824	1.359	0.670	1.581

Layer number	8	9	10	11	12	13	14	15	16
Indices (I)	31.62	80.48	189.9	385.3	635.6	846.6	958.6	994.3	1000.
Index ratio (I)	2.545	2.545	2.359	2.029	1.650	1.332	1.132	1.037	1.006
Indices (II)	0.621	1.581	0.670	1.359	1.824	1.098	0.970	1.006	1.000

TABLE 3.10 Indices of Refraction of Maximally Flat and Equal Ripple Designs†

R_{max} =	0%	0.1%	0.5%	1%	2%
n_0 =	1.000	1.000	1.000	1.000	1.000
n_1 =	2.234	3.128	3.508	3.743	4.048
n_2 =	0.179	0.180	0.190	0.198	0.209
n_3 =	8.309	6.986	6.903	6.931	7.030
n_4 =	0.108	0.147	0.163	0.173	0.186

$$n_5 = n_3 \quad n_6 = n_2 \quad n_7 = n_1 \quad n_s = n_0$$

†Have equal substrate index but different maximum reflectance in the passband.

The dotted curve of Fig. 3.10 gives the equal ripple corresponding to the equal ripple antireflection coating (solid curve).

We would like to emphasize that the reflectance of the equal ripple antireflection coating at φ *equals exactly* the reflectance of the equal ripple filter at 180° + φ. There are no approximations involved.

3.2.3. Riblet's procedure and equal ripple prototype filters

In Sec. 3.2.1 we presented data for equal ripple and maximally flat designs with two, three, and four layers using Collin's formulas (Table

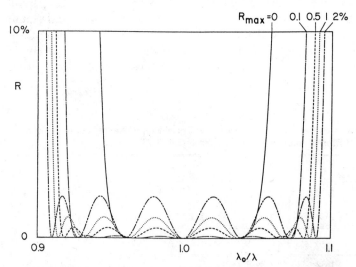

Figure 3.13. Reflectance in the passband of a maximally flat filter (R_{max} = 0, solid curve) and several equal ripple filters with different R_{max} [R_{max} = 0.1% (dash-dotted curve), = 0.5% (dashed curve), = 1 % (dotted curve), and = 2% (dash-dot-dotted curve)]. Designs are given in Table 3.10.

TABLE 3.11   Equal Ripple Prototype Filters for the Visual, R_{max} = 1% in the Passband

$m =$	15	17	19	21	23	25	27	29	31
$n_0 =$	1.674	1.676	1.671	1.665	1.671	1.670	1.670	1.669	1.668
$n_1 =$	2.003	1.984	1.986	1.978	1.977	1.970	1.968	1.968	1.966
$n_2 =$	1.644	1.653	1.663	1.666	1.675	1.678	1.682	1.684	1.686
$n_3 =$	2.162	2.129	2.122	2.107	2.100	2.089	2.083	2.080	2.076
$n_4 =$	1.533	1.547	1.562	1.569	1.580	1.586	1.592	1.595	1.599
$n_5 =$	2.289	2.253	2.240	2.221	2.212	2.197	2.189	2.184	2.178
$n_6 =$	1.470	1.480	1.495	1.502	1.513	1.519	1.526	1.530	1.534
$n_7 =$	2.353	2.326	2.316	2.297	2.289	2.274	2.265	2.260	2.253
$n_8 =$	1.450	1.450	1.460	1.466	1.475	1.480	1.486	1.490	1.494
$n_9 =$		2.349	2.350	2.337	2.332	2.319	2.311	2.307	2.300
$n_{10} =$			1.450	1.450	1.456	1.459	1.464	1.467	1.471
$n_{11} =$				2.349	2.351	2.342	2.336	2.333	2.327
$n_{12} =$					1.450	1.450	1.453	1.455	1.458
$n_{13} =$						2.348	2.347	2.346	2.341
$n_{14} =$							1.450	1.450	1.452
$n_{15} =$								2.350	2.348
$n_{16} =$									1.450

$$n_k = n_{m-k} \qquad n_s = n_0$$

3.4). Tables for maximally flat designs with up to eight layers were published by Young,[13] and tables for equal ripple designs with up to 21 layers were published by Levy.[15] The last two authors used Riblet's procedure (Riblet[16]) for the computations.

Riblet's procedure consists of the following steps:

1. Subject Eq. 3.24 to the transformation (Richards[17]):

$$s = \frac{i \sin \phi}{\cos \phi} \qquad (3.39)$$

2. Using the theorems of analytic continuation, determine the amplitude reflectance $\vec{R}$ from R (Eq. 3.24) by finding all the roots of the denominator $D = 1 + CT_m^2(s)$, excluding all the roots with positive real parts, and reconstructing D from

$$D = (s - s_1)(s - s_2)(s - s_3) \cdots (s - s_m)$$

$$= a_0 s^m + a_1 s^{m-1} + a_2 s^{m-2} + \cdots + a_{m-1} s^1 + a_m \qquad (3.40)$$

where $s_1, s_2, \ldots$ are the roots with negative real parts.

$$\vec{R} = \rho \frac{T_m(s)}{D(s)} \qquad (3.41)$$

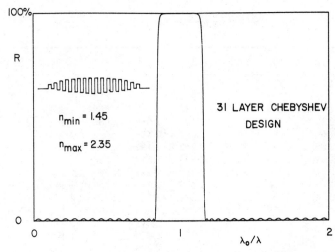

Figure 3.14. Reflectance of the 31-layer design given in Table 3.11.

3. Retransform $\vec{R}(s)$ to $\vec{R}(\cos \phi)$.

4. Solve for the elements of the characteristic matrix with the help of Eqs. 2.23 and 2.37.

5. Derive the final equations for the refractive indices of the individual layers by multiplying in sequence the characteristic matrix of the remaining multilayer from the left by the inverse of the characteristic matrix of each layer:

$$\begin{bmatrix} \cos \phi & \dfrac{-i \sin \phi}{n} \\ -in_k \sin \phi & \cos \phi \end{bmatrix}$$

$$\times \begin{bmatrix} a_0 \cos^{m-k+1}\phi + a_2 \cos^{m-k-1}\phi + \cdots \\ i \sin \phi(a_1 \cos^{m-k}\phi + a_3 \cos^{m-k-2}\phi + \cdots) \\ i \sin \phi(b_1 \cos^{m-k}\phi + b_3 \cos^{m-k-2}\phi + \cdots) \\ b_1 \cos^{m-k+1}\phi + b_2 \cos^{-k-1}\phi + \cdots) \end{bmatrix} \quad (3.42)$$

n_k is determined by setting the coefficients of $\cos^{m-k+2}\phi$ and $\sin \phi$ $\cos^{m-k+1}\phi$ equal to 0.

Example 1. Recently, a seven-layer maximally flat prototype filter was used to design a bandpass filter (Baumeister[18]). In Table 3.10 and Fig. 3.13 (both on p. 66) we compare this design ($R_{\max} = 0$) with designs having equal rejection ($n_S/n_0 = 10^9$ or $T_{\min} = 4 \times 10^{-9}$) but different $R_{\max}$.

Example. 2. In Table 3.11 equal ripple filters with different numbers of layers but equal maximum passband ripple ($R_{\max} = 1\%$) and equal maximum and min-

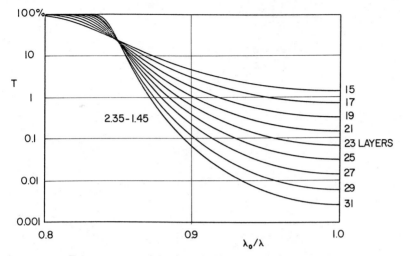

Figure 3.15. Transmittance of the designs given in Table 3.11.

imum indices of refraction are presented ($n_{max} = 2.35$ and $n_{min} = 1.45$). Figure 3.14 gives the reflectance of the design with the highest number of layers and Fig. 3.15 gives the transmittance of all the designs in the rejection band. These designs use refractive indices which are available in the visual region of the spectrum.

TABLE 3.12   Equal Ripple Prototype Designs for the Infrared, $R_{max} = 1\%$ in the Passband

Refractive index	17-layer 4.2–1.85	21-layer 4.2–1.85	21-layer 4.2–2.2	25-layer 4.2–2.2
$n_0 =$	2.519	2.519	2.749	2.748
$n_1 =$	3.167	3.154	3.336	3.327
$n_2 =$	2.260	2.281	2.630	2.645
$n_3 =$	3.682	3.641	3.682	3.656
$n_4 =$	1.997	2.024	2.406	2.427
$n_5 =$	4.021	3.974	3.958	3.923
$n_6 =$	1.886	1.909	2.280	2.300
$n_7 =$	4.163	4.125	4.112	4.079
$n_8 =$	1.850	1.865	2.222	2.239
$n_9 =$	4.199	4.184	4.181	4.155
$n_{10} =$		1.850	2.20	2.211
$n_{11} =$		4.200	4.200	4.189
$n_{12} =$				2.200
$n_{13} =$				4.199

$$n_k = n_{m-k} \qquad n_s = n_0$$

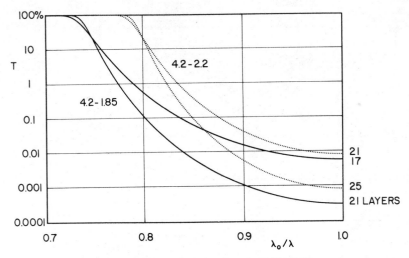

Figure 3.16. Transmittance of the designs given in Table 3.12.

Example 3. In Table 3.12 and Fig. 3.16 additional prototype designs, but now with different ratios of the maximum and minimum index, are given. These designs use refractive indices which are available in the infrared wavelength region.

3.2.4. Chebyshev filters with $R_{min} > 0$

In Eq. 3.22, if we set $P(\cos \phi) = 1 - C + \rho T_m(\cos \phi)$ we obtain

$$R = \frac{1 - C + T_m(\cos \rho)}{1 + T_m(\cos \rho)}$$

$$= R_{min} \frac{[1 + \rho T_m (\cos \phi)/R_{min}]}{1 + \rho T_m (\cos \phi)} \qquad (3.43)$$

Riblet's procedure (Sec. 3.2.3) can be used again to find the corresponding refractive index configuration. Now the reconstruction of the polynomial from its roots with negative real part has to be done for both numerator and denominator. Table 3.13 and Fig. 3.17 give examples with 1 percent ripple and various R_{min}.

3.2.5. Synthesis with elliptic functions

While maximally flat filters have two free parameters (number of layers and bandwidth), Chebyshev filters have three (number of layers, bandwidth, and ripple). In electrical network theory (Chen[19]) a third type of filter characteristic using Jacobian elliptic functions is often

TABLE 3.13   Chebyshev Designs with $R_{min} > 0$

	R_{min} (type curve)			
Refractive index	0.0 (solid)	0.1 (dashed)	0.2 (dotted)	0.3 (dash-dotted)
n_0	1.000	1.000	1.000	1.000
n_1	1.204	2.342	3.200	4.180
n_2	0.968	1.571	2.116	2.762
n_3	1.315	2.482	3.361	4.367
n_4	0.894	1.512	2.053	2.692
n_5	1.402	2.535	3.408	4.410
n_6	0.854	1.503	2.054	2.704
n_7	1.444	1.516	3.361	4.333
n_8	0.842	1.534	2.110	2.788
n_9	1.444	2.432	3.230	4.150
n_{10}	0.854	1.610	2.224	2.945
n_{11}	1.402	2.288	3.031	3.892
n_{12}	0.894	1.728	2.385	3.153
n_{13}	1.315	2.123	2.826	3.646
n_{14}	0.968	1.854	2.539	3.335
n_{15}	1.204	2.002	2.691	3.494
n_s	1.000	1.925	2.618	3.422

used. It has a fourth free parameter: the steepness of the edge k. For $k = 0$ the elliptic response is identical to the corresponding Chebyshev response. For $k > 0$ the response has equal ripple characteristic both in the stop band and in the passband. Unfortunately, the Jacobian elliptic functions can only be transformed into a ratio of two polyno-

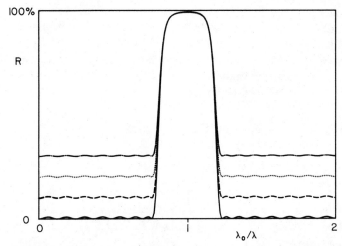

Figure 3.17. Reflectance of Chebyshev designs with 1 percent ripple and various minimum reflectances (see Table 3.13).

TABLE 3.14 Semielliptic 15-Layer Designs

Refractive index	k (type curve)			
	0 (solid)	0.25 (dash-dotted)	0.5 (dotted)	0.75 (dashed)
n_0	1	1	1	1
n_1	1.584	1.527	1.374	1.527
n_2	1.125	1.136	1.160	1.136
n_3	1.718	1.701	1.629	1.701
n_4	1.057	1.066	1.106	1.066
n_5	1.795	1.783	1.724	1.783
n_6	1.026	1.033	1.065	1.033
n_7	1.828	1.815	1.767	1.815
n_8	1.017	1.024	1.052	1.024

$$n_k = n_{m-k} \qquad n_s = n_0$$

mials. Consequently the necessary shape of Eq. 3.22 can not be simulated. It is possible, though, to go through the process of designing the filter as described by Chen[19] but ignoring the denominator polynomial. Figure 3.18 gives some examples. The parameter k allows a trade-off between ripple height near the edge and steepness of the edge slope.

3.3. Effective interfaces

In Sec. 2.9 we derive the split filter formula (Eq. 2.55) which allows us to study the transmittance of a multilayer as a function of the parameters of a layer inside the multilayer. This formula can also be used to study the combination of two known subsystems, A and B, and the effect of varying the parameters of the layer separating the two subsystems. Good use can be made of the fact that in Eq. 2.55 phases and amplitudes can be treated separately:

1. $T_0 = T_A T_B/(1 - r)^2$ depends on the amplitudes of reflectance and transmittance of the subsystems only.
2. T/T_0 depends primarily on the phases upon reflection and the phase thickness of the spacer.

Both parts are always ≤ 1. For high transmittance the amplitudes have to be adjusted so that $T_0 \approx 1$ and the phases so that $T/T_0 \approx 1$. These are the only possible solutions.

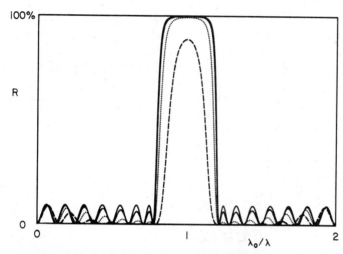

Figure 3.18. Reflectance of semielliptic 15-layer designs with $\rho = 0.1$, $W = 1.574$, and $k = 0, 0.25, 0.5, 0.75$ (see Table 3.14).

Out of the structure of Eq. 2.55 it follows that

$$T_0 = 1 \quad \text{only when} \quad R_A = R_B \tag{3.44}$$

and

$$\frac{T}{T_0} = 1 \quad \text{only when} \quad r = 0$$

or

$$\sin^2 \frac{\Phi_A + \Phi_B - 2\phi}{2} = 0 \tag{3.45}$$

Conversely,

$$T_0 = 0 \quad \text{the more } R_A \text{ and } R_B \text{ differ} \tag{3.46}$$

and

$$\frac{T}{T_0} = 0 \quad \text{the larger } r \text{ and } \sin^2 [\Phi_A + \Phi_B - 2\phi)/2] \text{ are} \tag{3.47}$$

In the design method of effective interfaces, Eqs. 3.44 to 3.47 are used as guides to combine two known multilayers into the desired filter (Smith[20]).

3.4. Absentee layers

In Prob. 2.3 we saw that the characteristic matrix of multiples of a half-wave-thick layer is plus or minus the unity matrix and no longer depends on the refractive index. The transmittance and reflectance of a multilayer are consequently not changed whenever we add or eliminate a half-wave layer. This is, of course, true only at exactly the wavenumber where the considered film is one-half wavelength thick. By adding a half-wave-thick layer, we can therefore change the transmittance and reflectance of a multilayer in the wavenumber region around a reference point without affecting the transmittance and reflectance at the reference point.

Often, the elimination of one half-wave layer generates another—which again can be eliminated. The following six designs all have equal transmittance and reflectance at the half-wave position:

$$H\ L\ H\ L\ H\ LL\ H\ L\ H\ L$$
$$H\ L\ H\ L\ HH\ L\ H\ L$$
$$H\ L\ H\ LL\ H\ L$$
$$H\ L\ HH\ L$$
$$H\ LL$$
$$H$$

3.5. Buffer Layers

Let us again look at the split filter equation (Eq. 2.55). Whenever r_A or $R_B = 0$ then $r = 0$ and $T = T_B$ or T_A. The transmittance of the multilayer is consequently no longer a function of the thickness of the spacer. Or in words: If inside a layer of a multilayer the reflectance to one side equals zero then thickness changes of this layer do not cause any variations in the transmittance and reflectance of the total multilayer (Mouchart,[21] Knittl[22]).

Buffer layers allow keeping the transmittance of a multilayer in one plane of polarization constant while changing it in the other. They also are very useful in the design of antireflection coatings.

3.6. Problems and Solutions

Problem 3.1

Determine the equivalent index of the three-layer configuration $n_1 = 2.3$, $QWOT_1 = 173.2$; $n_2 = 1.63$, $QWOT_2 = 193.2$; $n_3 = 2.3$, $QWOT_3 = 173.2$ at $\lambda = 540$ nm.

Solution. With Eqs. 2.5 and 2.56

$$\phi_1 = \frac{90°QWOT_1}{\lambda} = 28.87° \qquad \phi_2 = 16.1°$$

With Eqs. 2.17 and 2.20

$$\begin{bmatrix} 0.8758 & 0.3507i \\ 1.1104i & 0.8758 \end{bmatrix} \begin{bmatrix} 0.9608 & 0.1701i \\ 0.4520i & 0.9608 \end{bmatrix} = \begin{bmatrix} 0.7465 & 0.2507i \\ 1.4627i & 0.6525 \end{bmatrix}$$

We can avoid another matrix multiplication by using Eq. 2.28 together with Eqs. 3.1 and 3.2:

$$N = \sqrt{\frac{M_{21}M_{22}}{M_{12}M_{11}}} = 1.909$$

$$\Gamma = \sin^{-1}\frac{2M_{21}M_{22}}{N} = 89.13°$$

Problem 3.2

Derive a general formula for the equivalent index of a sequence of layers for large wavelengths λ (small wavenumbers).

Solution. Let the sequence of layers be described by n_1, $QWOT_1$; n_2, $QWOT_2$; n_3, $QWOT_3$; . . . . For small λ, Eq. 2.17 takes the form

$$\mathbf{M} = \begin{bmatrix} 1 & \dfrac{i\delta}{n} \\ in\delta & 1 \end{bmatrix} \tag{3.48}$$

With Eq. 3.48 the characteristic matrix of the sequence of layers becomes

$$\mathbf{M} = \begin{bmatrix} 1 & i(\delta_1/n_1 + \delta_2/n_2 + \delta_3/n_3 + \cdots) \\ i(n_1\delta_1 + n_2\delta_2 + n_3\delta_3 + \cdots) & 1 \end{bmatrix} \tag{3.49}$$

which yields for the equivalent index N

$$N = \sqrt{\frac{n_1\delta_1 + n_2\delta_2 + n_3\delta_3 + \cdots}{n_1/\delta_1 + n_2/\delta_2 + n_3/\delta_3 + \cdots}} \tag{3.50}$$

or after division by $90°/\lambda$

$$N = \sqrt{\frac{n_1 QWOT_1 + n_2 QWOT_2 + n_3 QWOT_3 + \cdots}{n_1/QWOT_1 + n_2/QWOT_2 + \cdots}} \tag{3.51}$$

Problem 3.3

Apply the general formula derived in Prob. 3.2 for the two particular cases H/2 L H/2 and L/2 H L/2.

Solution.

$$N_{H/2\,L\,H/2} = \sqrt{\frac{n_H + n_L}{1/n_H + 1/n_L}} = \sqrt{n_H n_L} = N_{L/2\,H\,L/2} \qquad (3.52)$$

in both cases.

Problem 3.4

Determine the equivalent index of a quarter-wave stack with an odd number of layers. Assume all layers to have different refractive indices.

Solution.   With Eq. 2.18

$$
\mathbf{M} = \begin{bmatrix} 0 & \dfrac{i}{n_1} \\ in_1 & 0 \end{bmatrix} \begin{bmatrix} 0 & \dfrac{i}{n_2} \\ in_2 & 0 \end{bmatrix} \begin{bmatrix} 0 & \dfrac{i}{n_3} \\ in_3 & 0 \end{bmatrix} \cdots \begin{bmatrix} 0 & \dfrac{i}{n_m} \\ in_m & 0 \end{bmatrix}
$$

$$
= \begin{bmatrix} \dfrac{-n_2}{n_1} & 0 \\ 0 & \dfrac{-n_1}{n_2} \end{bmatrix} \begin{bmatrix} \dfrac{-n_4}{n_3} & 0 \\ 0 & \dfrac{-n_4}{n_3} \end{bmatrix} \cdots \begin{bmatrix} 0 & \dfrac{i}{n_m} \\ in_m & 0 \end{bmatrix}
$$

$$
= \begin{bmatrix} -\dfrac{n_2 n_4 n_6 \cdots}{n_1 n_3 n_5 \cdots} & 0 \\ 0 & \pm\dfrac{n_1 n_3 n_5 \cdots}{n_2 n_4 n_6 \cdots} \end{bmatrix} \begin{bmatrix} 0 & \dfrac{i}{n_m} \\ in_m & 0 \end{bmatrix}
$$

$$
= \begin{bmatrix} 0 & \pm i\,\dfrac{n_2 n_4 n_6 \cdots}{n_1 n_3 n_5 \cdots} \\ \pm i\,\dfrac{n_1 n_3 n_5 \cdots}{n_2 n_4 n_6 \cdots} & 0 \end{bmatrix} \qquad (3.53)
$$

and, using Eq. 3.1,

$$N = \sqrt{\frac{M_{21}}{M_{12}}} = \frac{n_1 n_3 n_5 \cdots n_m}{n_2 n_4 n_6 \cdots n_{m-1}} \qquad (3.54)$$

Since both matrix elements M_{11} and M_{22} of matrix equation 3.53 are zero they are also equal. This means that an odd-numbered quarter-wave stack has an equivalent index at the quarter-wave position whether it is symmetric or not!

In addition, since N does not change its value when indices of layers are exchanged as long as they stay on odd- or even-numbered places, it is invariant to reversal of direction.

Problem 3.5

Write a Basic program to calculate equal ripple configurations after
Riblet,[16] Young,[14] and Levy.[15]

Solution.

Program Listing for Problem 3.5

```
100    ! NAME IS HP9816-"RIBLET"
101    ! CALCULATES EQUAL RIPPLE PROTOTYPE FILTERS
102    OPTION BASE 0
103    DIM Num1(52),Num2(52),Denom1(52),Denom2(52)
104    DIM Index(52),Swr(52),Rroot1(52),Iroot1(52)
105    DIM Rroot2(52),Iroot2(52),Zero(52),Hindex(52)
106    DEG
107    DATA 15,1.5,1000
108    FOR I=0 TO 52
109      Zero(I)=0
110    NEXT I
111    READ Dim,Bandw,Subindex
112    PRINT
113    PRINT "EQUAL RIPPLE PROTOTYPE FILTER CALCULATION AFTER RIBLET"
114    PRINT
115    PRINT Dim;"LAYERS";"    ";"W=";Bandw;"    ";"NS/NO=";Subindex
116    PRINT
117    REDIM Denom1(Dim),Denom2(Dim),Zero(Dim),Rroot1(Dim),Iroot1(Dim)
118    REDIM Rroot2(Dim),Iroot2(Dim),Num1(Dim)
119    MAT Num1= Zero
120    Num1(0)=1
121    FOR I=2 TO Dim STEP 2      ! CALCULATE CHEBYSHEV COEFFICIENTS
122      IF I=Dim THEN
123        X=2
124        GOTO 127
125      END IF
126      X=2*FNBinomi(Dim-I/2,I/2)-FNBinomi(Dim-I/2-1,I/2)
127      Num1(I)=(-1)^(I/2)*INT(X+.2)
128    NEXT I
129    M=1/SIN(45*Bandw)
130    M1=4*M*M-2                 ! Tn(2*x^2-1)=2*Tn^2(x)-1
131    K1=Num1(0)
132    FOR I=2 TO Dim STEP 2
133      K1=K1*M1*M1+Num1(I)
134    NEXT I
135    IF Dim/2-INT(Dim/2)<>0 THEN K1=K1*M1
```

Program Listing for Problem 3.5 *(Continued)*

```
136    K1=(Subindex-1)*(Subindex-1)/(K1+2)/Subindex
137    PRINT "RIPPLE%=";100*K1
138    PRINT
139    PRINT " #            n_ERAR           n_ERF         R%"
140    MAT Rroot1= Zero
141    MAT Iroot1= Zero
142    H1=1/SQR(K1)
143    H1=LOG(H1+SQR(1+H1*H1))/Dim
144    FOR I=0 TO Dim-1
145       Rroot1(I)=-.5*(EXP(H1)-EXP(-H1))*SIN((2*I+1)*180/2/Dim)
146       Iroot1(I)=.5*(EXP(H1)+EXP(-H1))*COS((2*I+1)*180/2/Dim)
147    NEXT I
148    FOR J=0 TO Dim-1
149       CALL Cmult(Rroot1(J),Iroot1(J),Rroot1(J),Iroot1(J),H1,H2)
150       CALL Cdivid(M*M,0,H1,H2,H3,H4)
151       H3=1+H3
152       CALL Csqrt(H3,H4,H1,H2)
153       Rroot1(J)=H1
154       Iroot1(J)=H2
155    NEXT J
156    MAT Rroot2= Rroot1
157    REDIM Rroot2(Dim-1)
158    MAT SORT Rroot2
159    FOR J=0 TO Dim-1 STEP 2
160       FOR I=0 TO Dim-1
161          IF Rroot2(J)=Rroot1(I) THEN
162             Iroot2(J)=Iroot1(I)
163             Iroot2(J+1)=-Iroot1(I)
164             GOTO 167
165          END IF
166       NEXT I
167    NEXT J
168    REDIM Denom1(Dim+1),Denom2(Dim+1),Zero(Dim+1)
169    MAT Denom1= Zero
170    Denom1(1)=1
171    MAT Denom2= Denom1
172    F=0
173    FOR I=0 TO Dim-1
174       IF ABS(Iroot2(I))<1.E-6 THEN
175          GOSUB 258
176          GOTO 179
177       END IF
178       GOSUB 264
```

Program Listing for Problem 3.5 *(Continued)*

```
179  NEXT I
180  REDIM Denom2(Dim)
181  FOR I=0 TO Dim
182     Denom2(I)=Denom1(Dim-I+1)
183  NEXT I
184  REDIM Num2(Dim),Zero(Dim)
185  MAT Num2= Zero
186  M1=2*M
187  MAT Num1= (.5)*Num1
188  FOR I=0 TO Dim STEP 2
189     Num2(I)=Num1(I)*(-1)^(I/2)
190     IF I=Dim THEN 195
191     FOR K=I+2 TO Dim STEP 2
192        Num2(I)=Num2(I)*M1*M1+(-1)^(I/2)*Num1(K)*FNBinomi(K/2,I/2)
193     NEXT K
194     IF Dim/2-INT(Dim/2)<>0 THEN Num2(I)=Num2(I)*M1
195  NEXT I
196  K2=(1-Subindex)*Denom2(0)/(1+Subindex)/Num2(0)
197  MAT Num2= (K2)*Num2
198  K1=0
199  K2=0
200  FOR I=0 TO Dim
201     K3=Denom2(I)-Num2(I)
202     K4=Denom2(I)+Num2(I)
203     Denom2(I)=K3
204     Num2(I)=K4
205     K1=K1+K3
206     K2=K2+K4
207  NEXT I
208  Index(1)=K1/K2
209  FOR J=Dim TO 1 STEP -1
210     REDIM Num1(J+1),Denom1(J+1)
211     Num1(0)=-Denom2(0)
212     Num1(J+1)=Index(Dim-J+1)*Num2(J)
213     Denom1(0)=-Index(Dim-J+1)*Num2(0)
214     Denom1(J+1)=Denom2(J)
215     FOR I=1 TO J
216        Num1(I)=Index(Dim-J+1)*Num2(I-1)-Denom2(I)
217        Denom1(I)=Denom2(I-1)-Index(Dim-J+1)*Num2(I)
218     NEXT I
219     REDIM Denom2(J-1),Num2(J-1)
220     MAT Num1= (Index(Dim-J+1))*Num1
221     MAT Denom2= Zero
```

Program Listing for Problem 3.5 *(Continued)*

```
222    MAT Num2= Zero
223    K1=0
224    K2=0
225    FOR I=J-1 TO 0 STEP -2
226       FOR K=J+1 TO I+2 STEP -2
227          Denom2(I)=Denom2(I)+Num1(K)
228          Num2(I)=Num2(I)+Denom1(K)
229       NEXT K
230       K1=K1+Denom2(I)
231       K2=K2+Num2(I)
232    NEXT I
233    FOR I=J-2 TO 0 STEP -2
234       FOR K=J TO I+2 STEP -2
235          Denom2(I)=Denom2(I)+Num1(K)
236          Num2(I)=Num2(I)+Denom1(K)
237       NEXT K
238       K1=K1+Denom2(I)
239       K2=K2+Num2(I)
240    NEXT I
241    Swr(Dim-J+2)=K1/K2/Index(Dim-J+1)
242    Index(Dim-J+2)=K1/K2
243    NEXT J
244    Swr(0)=1
245    Swr(1)=Index(1)
246    Index(0)=1
247    Hindex(0)=1
248    FOR I=1 TO Dim+1
249       J=(-1)^(I+1)
250       Hindex(I)=Hindex(I-1)*Swr(I)^J
251    NEXT I
252    FOR I=0 TO Dim+1
253       Refl=100*((1-Swr(I))/(1+Swr(I)))^2
254       PRINT I,Index(I);TAB(30);Hindex(I);TAB(50);Refl
255    NEXT I
256    PRINT
257    STOP
258    REM*******
259    FOR L=2 TO Dim+1
260       Denom2(L)=Denom1(L)+Rroot2(I)*Denom1(L-1)
261    NEXT L
262    MAT Denom1= Denom2
263    RETURN
264    REM*******
```

Program Listing for Problem 3.5   *(Continued)*

```
265    IF F=1 THEN
266      F=0
267      RETURN
268    END IF
269    F=1
270    A1=-2*Rroot2(I)
271    A2=Rroot2(I)^2+Iroot2(I)^2
272    FOR L=2 TO Dim+1
273      Denom2(L)=Denom1(L)-A1*Denom1(L-1)+A2*Denom1(L-2)
274    NEXT L
275    MAT Denom1= Denom2
276    RETURN
277    END
278    DEF FNBinomi(X1,X2)! CALCULATES BINOMIAL COEFFICIENT (X1 OVER X2)
279      X=1
280      FOR I=X1 TO X1-X2+1 STEP -1
281        X=I*X
282      NEXT I
283      FOR I=1 TO X2
284        X=X/I
285      NEXT I
286      RETURN X
287    FNEND
288    SUB Cadd(N,A_(*),B_(*),Real,Imag)
289      Real=0
290      Imag=0
291      FOR I=1 TO N
292        Real=Real+A_(I)
293        Imag=Imag+B_(I)
294      NEXT I
295    SUBEND!

296    SUB Cmult(A1,B1,A2,B2,R,I)
297      R=A1*A2-B2*B1
298      I=A1*B2+B1*A2
299    SUBEND!

300    SUB Cdivid(A1,B1,A2,B2,R,I)
301      C=A2*A2+B2*B2
302      IF C<>0 THEN 306
303      PRINT FNLin$(2);"ERROR IN SUBPROGRAM Cdivid."
304      PRINT "DIVISOR IS ZERO.",FNLin$(2)
305      PAUSE
```

Program Listing for Problem 3.5   *(Continued)*

```
306     R=(A2*A1+B2*B1)/C
307     I=(A2*B1-B2*A1)/C
308   SUBEND!


309   SUB Cexp(A,B,R,I)
310     RAD
311     R=EXP(A)*COS(B)
312     I=EXP(A)*SIN(B)
313   SUBEND!


314   SUB Clog(A,B,R,I)
315     RAD
316     IF A<>0 OR B<>0 THEN 320
317     PRINT FNLin$(2);"ERROR IN SUBPROGRAM Clog."
318     PRINT "UNDEFINED FUNCTION FOR A=B=0",FNLin$(2)
319     PAUSE
320     R=LOG(SQR(A*A+B*B))
321     IF A=0 THEN
322        I=ASN(SGN(B))
323     ELSE
324        I=ATN(B/A)+ACS(-1)*(A<0)*(SGN(B)+(B=0))
325     END IF
326   SUBEND!


327   SUB Polyev(N,A,B,Rcoef(*),Icoef(*),Rval,Ival)
328     Rval=Rcoef(N)
329     Ival=Icoef(N)
330     FOR I=N-1 TO 0 STEP -1
331        T=Rval*A-Ival*B+Rcoef(I)
332        Ival=Rval*B+Ival*A+Icoef(I)
333        Rval=T
334     NEXT I
335   SUBEND!


336   SUB Cabs(X,Y,Cabs)
337     X1=ABS(X)
338     Y1=ABS(Y)
339     IF X1<>0 THEN 342
340     Cabs=Y1
341     SUBEXIT
342     IF Y1<>0 THEN 345
343     Cabs=X1
344     SUBEXIT
```

Program Listing for Problem 3.5   *(Continued)*

```
345     IF X1>Y1 THEN Cabs=X1*SQR(1+(Y1/X1)^2)
346     IF X1<=Y1 THEN Cabs=Y1*SQR(1+(X1/Y1)^2)
347  SUBEND!


348  SUB Csqrt(A,B,R,I)
349     IF (A<>0) OR (B<>0) THEN 353
350     R=0
351     I=0
352     SUBEXIT
353     CALL Cabs(A,B,Cabs)
354     R=SQR((ABS(A)+Cabs)*.5)
355     IF A<0 THEN 358
356     I=B/(R+R)
357     SUBEXIT
358     IF B<0 THEN I=-R
359     IF B>=0 THEN I=R
360     R=B/(I+I)
361  SUBEND!
```

Sample Calculation for Problem 3.5

```
EQUAL RIPPLE PROTOTYPE FILTER CALCULATION AFTER RIBLET

 15 LAYERS    W= 1.5    NS/NO= 1000

RIPPLE%= .557062256505
```

#	n_ERAR	n_ERF	R%
0	1	1	0
1	1.17471242124	1.17471242124	.645422193134
2	1.4745113238	.935868887774	1.2806274291
3	2.06103544242	1.30813437376	2.75206431284
4	3.17766386911	.848457048582	4.54328955191
5	5.29848268796	1.41472955245	6.2605173324
6	9.34056976156	.802512077223	7.62404403226
7	17.0462838377	1.46456254819	8.52806625154
8	31.6227766016	.789473650874	8.97015935329
9	58.6638125655	1.46456254819	8.97015935327
10	107.059850257	.802512077224	8.52806625149
11	188.733276842	1.41472955245	7.62404403218
12	314.6965951	.848457048584	6.26051733231
13	485.19301483	1.30813437376	4.54328955182

Sample Calculation for Problem 3.5 (Continued)

14	678.190790288	.935868887776	2.75206431279
15	851.272176826	1.17471242124	1.28062742908
16	999.999999972	1	.64542219313

The program uses standard Hewlett-Packard subroutines for the complex operations (Cmult, Cdivid, etc.). FNBinomi calculates binomial coefficients X1 over X2.

Antireflection Coatings

Antireflection coatings are used to reduce the (Fresnel) surface reflectance of optical components (against air), to reduce the reflectance of an interface between two massive optical media with different refractive indices (e.g., cemented lenses), to match a coating design optimized for one massive optical medium into another, or to match two coating designs optimized for different massive media into each other when they are combined into a compound filter.

The potential of antireflection coatings with a few layers can be read from their reflectance formulas. For more complex antireflection coatings, design models have to be used. The limitations of these models reduce, of course, the generality of the derived solutions.

It is relatively easy to eliminate the reflectance around one or two wavelength positions. Yet the reduction of the reflectance over an extended wavelength region is rather difficult, especially when the ratio of the wavelengths at the edges of the low-reflectance region exceeds two.

The ideal antireflection coating is an inhomogeneous layer with a continuous transition of the refraction index from one massive medium to the other or, as an approximation, a set of very thin homogeneous layers with refractive indices increasing in small steps from the low-index massive medium to the high-index medium. This coating is of no practical value since the choice of materials which can be deposited into hard and environmentally stable coatings is very limited, mixing of two materials to generate in-between refractive indices is very difficult, and, in the typical case when one of the massive media is air, the lowest available refractive index is 1.38 (MgF$_2$)—much too large a step from 1. To achieve low reflectance towards air ($n_0 = 1$) over as large a wavelength region as possible, while overcoming the basic limitation that there are no usable coating materials with a refractive index below 1.38, is one of the fundamental challenges of interference coating design.

4.1. Single-Layer Antireflection Coatings

For the transmittance of a single film on a substrate at normal incidence we obtain, from Eqs. 2.17 and 2.40,

$$T = \frac{4}{2 + n_0/n_s + n_s/n_0 + (n_s - n^2/n_s)(n_0/n^2 - 1/n_0)\sin^2\phi} \quad (4.1)$$

The extrema of this function are

$$T_{min} = \frac{4n_0n_s}{(n_0+n_s)^2} = 1 - R_{max} \quad \text{for } \phi = 0, 180°, 360°, \ldots \quad (4.2)$$

or
$$R_{max} = \frac{(n_s-n_0)^2}{(n_s+n_0)^2}$$

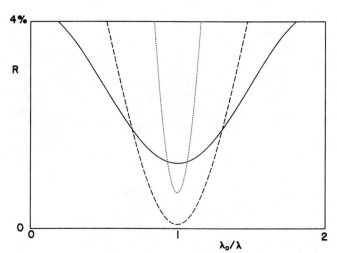

Figure 4.1. Reflectance of single-layer antireflection coatings: 1.0 | L | n_S with $n_{L1} = 1.38$ and $n_{S1} = 1.52$ (solid curve), $n_{L2} = 1.38$ and $n_{S2} = 1.75$ (dashed curve), $n_{L3} = 2.2$ and $n_{S3} = 4.1$ (dotted curve).

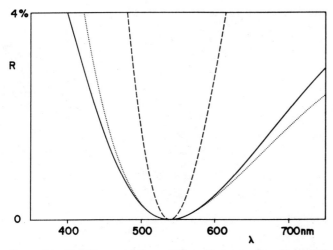

Figure 4.2. Reflectance of three "V"-coatings on glass ($\lambda_0 = 540$ nm): $1.0 \,|\, \text{L H} \,|\, 1.52$ with $n_{L1} = 1.38$ and $n_{H1} = 1.7$ (solid curve), with $n_{L2} = 1.86$ and $n_{H2} = 2.3$ (dashed curve); $1.0 \,|\, 1.29\text{L } 0.211\text{H} \,|\, 1.52$ with $n_{L3} = 1.38$ and $n_{H3} = 2.3$ (dotted curve).

(which is equal to the reflectance of the uncoated substrate) and

$$T_{\max} = \frac{4n^2 n_0 n_s}{(n_0 n_s + n^2)^2} = 1 - R_{\min} \qquad \text{for } \phi = 90°, 270°, \ldots \quad (4.4)$$

$$\text{or} \qquad R_{\min} = \frac{(n_0 n_s - n^2)^2}{(n_0 n_s + n^2)^2} \qquad (4.5)$$

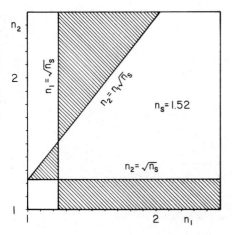

Figure 4.3. Index zones (shaded areas) where solutions to Eq. 4.8 are possible.

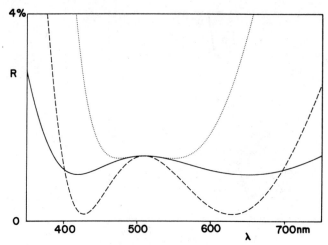

Figure 4.4. Reflectance of three "W"-coatings on glass ($\lambda_0 = 510$ nm): 1.0 | L HH | 1.52 with $n_L = 1.38$ and $n_{H1} = 1.6$ (solid curve), $n_{H2} = 2$ (dashed curve), and $n_{H3} = 2.5$ (dotted curve).

Consequently, $R_{min} = 0$ when

$$n = \sqrt{n_0 n_s} \quad \text{and} \quad nd = \frac{\lambda}{4}, \frac{3\lambda}{4}, \ldots \tag{4.6}$$

Figure 4.1 gives examples of various single-layer antireflection coatings on various substrates.

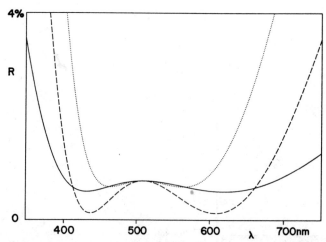

Figure 4.5. Reflectance of three "W"-coatings on glass with a higher refractive index than in Fig. 4.4 ($\lambda_0 = 510$ nm): 1.0 | L HH | 1.6 with $n_L = 1.38$ and $n_{H1} = 1.7$ (solid curve), $n_{H2} = 2.1$ (dashed curve), and $n_{H3} = 2.4$ (dotted curve).

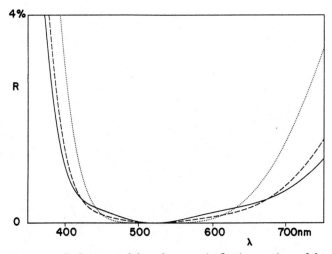

Figure 4.6. Reflectance of three-layer antireflection coatings of the $\lambda/4-\lambda/2-\lambda/4$ type ($\lambda_0 = 520$ nm): $1.0 \mid L\ HH\ M \mid 1.52$ with $n_L = 1.38$, $n_M = 1.70$, and $n_{H1} = 2.08$ (solid curve), $n_{H2} = 2.15$ (dashed curve), and $n_{H3} = 2.35$ (dotted curve).

4.2. Two-Layer Antireflection Coatings

For the transmittance of two layers on a substrate at normal incidence we obtain, from Eqs. 2.17, 2.22, and 2.40,

$$T = 4n_0n_s\{(n_0 + n_s)\cos\phi_1\cos\phi_2 - (n_0n_2/n_1 + n_sn_1/n_2)\sin\phi_1\sin\phi_2]^2$$

$$+ [(n_1 + n_0n_s/n_1)\sin\phi_1\cos\phi_2 + (n_2 + n_0n_s/n_2)\cos\phi_1\sin\phi_2]^2\}^{-1}$$

$$= 4n_0n_s\{4n_0n_s + [(n_0 - n_s)\cos\phi_1\cos\phi_2 - (n_0n_2/n_1 - n_sn_1/n_2)\sin\phi_1\sin\phi_2]^2$$

$$+ [(n_1 - n_0n_s/n_1)\sin\phi_1\cos\phi_2 + (n_2 - n_0n_s/n_2)\cos\phi_1\sin\phi_2]^2\}^{-1}$$

$$(4.7)$$

$T = 1$ or $R = 0$ when

$$\tan\phi_1\tan\phi_2 = \pm\frac{(n_0 - n_s)}{n_0n_2/n_1 - n_sn_1/n_2}$$

$$\frac{\tan\phi_1}{\tan\phi_2} = \pm\frac{(n_2 - n_0n_s/n_2)}{n_1 - n_0n_s/n_1}$$

$$(4.8)$$

or when

$$\frac{n_2}{n_1} = \sqrt{\frac{n_s}{n_0}} \quad \text{and} \quad \phi_1 = \phi_2 = \frac{\lambda_0}{4} \qquad (4.9)$$

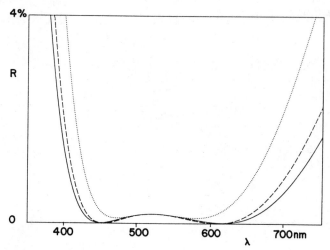

Figure 4.7. Reflectance of the same antireflection coatings as in
Fig. 4.6 but now $n_M = 1.63$.

Figure 4.2 gives examples. Figure 4.3 shows the index regions where
solutions of Eq. 4.8 are possible (Schuster[1]).

The antireflection coatings of Fig. 4.2 (often called "V"-coatings) do
provide zero reflectance but are very wavelength selective. A wide
region of low but not zero reflectance is provided by the $\lambda/4$–$\lambda/2$ coatings
(often called the "W"-coating). The $\lambda/4$ low-index layer provides the
low reflectance at λ_0. The $\lambda/2$ high-index layer, as "absentee" layer,
does not disturb the low reflectance at λ_0 but reverses the curvature

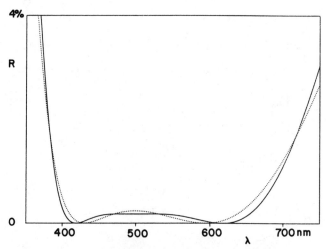

Figure 4.8. Three-layer antireflection coating of the $\lambda/4$–$\lambda/2$–$\lambda/2$
type (1.0 | L HH MM | 1.75 with $n_L = 1.38$, $n_H = 2.1$, and $n_M = 1.62$,
solid curve) compared to the $\lambda/4$–$\lambda/2$–$\lambda/4$ type on low-index glass
(1.0 | L HH M | 1.52, same indices, dotted curve) ($\lambda_0 = 500$ nm).

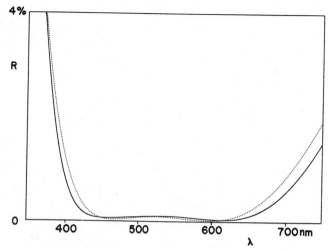

Figure 4.9. Reflectance of four-layer two-material antireflection coatings on two glass substrates ($\lambda_0 = 520$ nm): 1.0 | L 2.108H 0.331L 0.225H | 1.52 (solid curve) and 1.0 | L 2.108H 0.242L 0.219H | 1.64 (dotted curve) with $n_L = 1.38$ and $n_H = 2.08$.

of the reflectance curve of the single film (solid curve of Fig. 4.1) around λ_0 and generates broad low reflectance zones on both sides of λ_0. Figures 4.4 and 4.5 give examples.

4.3.   Three-Layer Antireflection Coatings on Glass

4.3.1.   The $\lambda/4–\lambda/2–\lambda/4$ coating†

With the success of broadening the low reflecting zone of the single layer by inserting a half-wave layer between the film and the substrate, it appears tempting to add a half-wave layer to the "V"-coating designs of Fig. 4.2. Figure 4.6 gives examples. By slightly deviating from Eq. 4.9 it is possible to improve the color balance of the residual reflection (Musset and Thelen,[3] Fig. 4.7).

4.3.2.   The $\lambda/4–\lambda/2–\lambda/2$ coating‡

At a substrate index 1.63 the bottom layer of the design in Fig. 4.7 is no longer effective. At higher substrate indices we can still use the design, though, if we insert a matching layer between the substrate and the coating. This leads to the $\lambda/4–\lambda/2–\lambda/2$ coating of Fig. 4.8. It turns out to be even wider than the $\lambda/4–\lambda/2–\lambda/4$ coating.

---

†From Geffcken.[2]
‡From Louderback and Zook.[4]

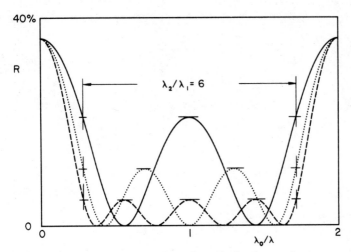

Figure 4.10. Reflectance of equal ripple antireflection coatings on germanium: $1.0 \mid L\ H \mid 4$ with $n_L = 1.809$ and $n_H = 2.211$ (solid curve); $1.0 \mid L\ M\ H \mid 4$ with $n_L = 1.579$, $n_M = 2$, and $n_H = 2.533$ (dotted curve); $1.0 \mid L\ M1\ M2\ H \mid 4$ with $n_L = 1.395$, $n_{M1} = 1.762$, $n_{M2} = 2.27$, and $n_H = 2.867$ (dashed curve). ($\lambda_2/\lambda_1 = 6$).

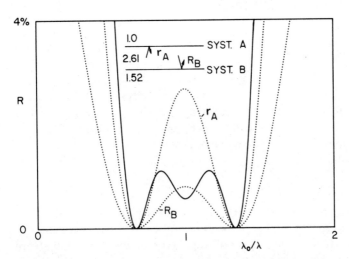

Figure 4.11. Reflectance of a step-up–step-down design: $1.0 \mid L\ M1\ H\ M2 \mid 1.52$ (solid curve) with $n_L = 1.38$, $n_{M1} = 1.89$, $n_{M2} = 1.82$, and $n_H = 2.18$. Also shown as dotted curves are the reflectances $2.61 \mid M1\ L \mid 1$ (labelled r_A) and $2.61 \mid H\ M2 \mid 1.52$ (labelled R_B).

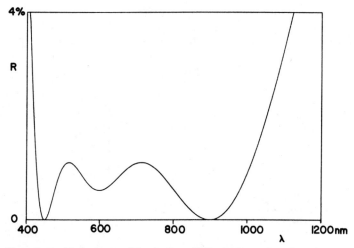

Figure 4.12. Reflectance of the design of Fig. 4.11 as a function of wavelength. (λ_0 = 600 nm).

TABLE 4.1 Adaptation of the Four-Layer Two-Material Antireflection Coating 1 | aL bH cL dH | n_s to Various Substrate Indices n_s†

n_s	1.45	1.48	1.50	1.52	1.56	1.60	1.62	1.64	1.68	1.72	1.74
c	0.385	0.365	0.350	0.331	0.300	0.271	0.250	0.242	0.213	0.185	0.188
d	0.223	0.227	0.227	0.225	0.229	0.225	0.223	0.219	0.208	0.188	0.175

†n_L = 1.38, n_H = 2.08, λ_0 = 520 nm, a = 1.0, and b = 2.108.

4.4. Four-Layer Two-Material Antireflection Coating†

There are two disadvantages to the $\lambda/4$–$\lambda/2$–$\lambda/4$ antireflection coating:

1. Ideal match requires a different index of the bottom layer for each new substrate index (Eq. 4.9).

2. Three coating materials are needed.

These deficiencies are overcome by the four-layer two-material antireflection coating. It is generated by replacing the bottom layer with an equivalent layer constructed from the other two materials (Sec. 3.1.3). Figure 4.9 gives examples for two different substrate indices. Table 4.1 gives the thicknesses for additional substrate indices. All thicknesses were numerically optimized through refining (Chap. 11).

---

†From Geffcken.[5]

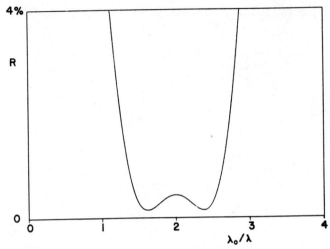

Figure 4.13. Reflectance of the coating $1.0 \mid L/2\,H/2 \mid 3.0$ with $n_L = 1.45$ and $nH = 2.35$.

4.5. Equal Ripple Antireflection Coating Synthesis

The one- to four-layer antireflection coatings discussed so far were discovered one at a time and then applied to whatever use one could find. Synthesis with Chebyshev polynomials (Sec. 3.2) takes the opposite approach: For each substrate index n_s, surrounding medium

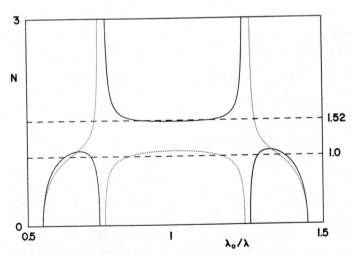

Figure 4.14. Equivalent indices of the two images of the $\lambda/4$–$\lambda/2$–$\lambda/4$ coating of Fig. 4.7 (solid curve): M HH LL HH M (solid curve) and L HH MM HH L (dotted curve) with $n_L = 1.38$, $n_M = 1.63$, and $n_H = 2.08$.

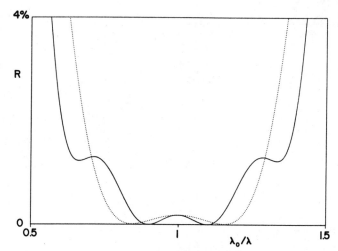

Figure 4.15. Reflectance   of   the   antireflection   coating
1.0 | L HH M LL | 1.52 (solid curve) compared to 1.0 | L HH M | 1.52
(dotted curve) (n_L = 1.38, n_M = 1.63, and n_H = 2.08).

index n_0, bandwidth λ_2/λ_1, and maximum residual reflectance R_{max} the
theory provides a minimum number of layers and their refractive in-
dices which will generate a perfect equal ripple characteristic. Yet
there are two major problems:

1. The refractive indices delivered by the theory can be any value

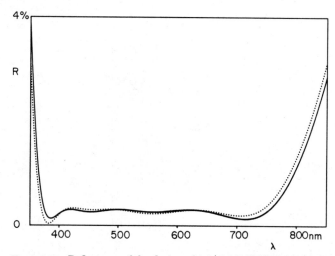

Figure 4.16.   Reflectance of the design of 1.0 | L M3 H M2 L M1 | 1.52
with n_L = 1.38, n_{M1} = 1.44, n_{M2} = 1.62, n_{M3} = 1.98, and n_H = 2.08
(solid curve). M3 can be synthesized with 0.46H 0.074L 0.46H (dot-
ted curve). (λ_0 = 500 nm).

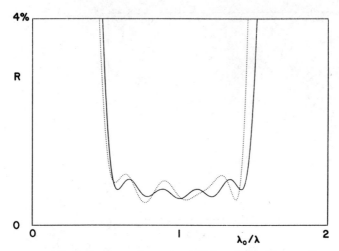

Figure 4.17. Reflectance of a seven-layer antireflection coating:
1.0 | L M3 M4 M2 M1 L M1 | 1.52 with $n_L = 1.38$, $n_{M1} = 1.45$,
$n_{M2} = 1.71$, $n_{M3} = 1.81$, and $n_{M4} = 1.95$ (solid curve). Synthesization with L and H ($n_H = 2.08$) gives for M2: 0.32L 0.34H 0.32L,
for M3: 0.26L 0.45H 0.26L, and for M4: 0.17L 0.63H 0.17L (dotted
curve).

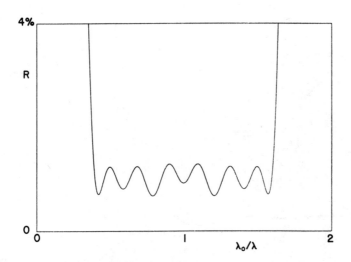

Figure 4.18. Reflectance of the nine-layer antireflection coating:
1.0 | L M3 M5 H M5 M4 M2 M1 M1 | 1.65 with $n_L = 1.38$, $n_{M1} = 1.58$,
$n_{M2} = 1.66$, $n_{M3} = 1.76$, and $n_{M4} = 1.86$, $n_{M5} = 2.14$, and $n_H = 2.28$.
One percent points have a wavelength ratio of 4.

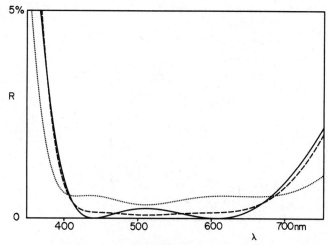

Figure 4.19. Reflectance of several triple-layer antireflection coatings of the classical quarter/half/quarter type ($\lambda_0 = 510$ nm): $1.0 \mid L1\ H1H1\ M1 \mid 1.52$ (solid curve), $1.0 \mid L2\ H2H2\ M2 \mid 1.52$ (dashed curve), and $1.0 \mid L2\ H1H1\ M3 \mid 1.52$ (dotted curve), with $n_{L1} = 1.38$, $n_{L2} = 1.46$, $n_{H1} = 2.08$, $n_{H2} = 2.35$, $n_{M1} = 1.62$, $n_{M2} = 1.75$, and $n_{M3} = 1.70$.

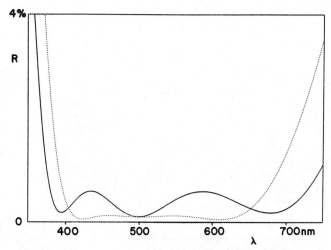

Figure 4.20. Reflectance of two antireflection coatings with $n_L = 1.45$ ($\lambda_0 = 500$ nm): $1.0 \mid L\ HH\ M\ L \mid 1.52$ with $n_M = 1.65$ and $n_H = 2.15$ (solid curve) and $1.0 \mid L\ H\ M2\ M1\ L \mid 1.52$ with $n_{M1} = 1.58$, $n_{M2} = 2.2$, and $n_H = 2.30$ (dotted curve).

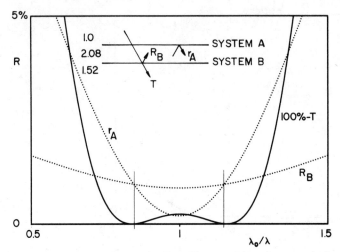

Figure 4.21.   Reflectance of $1 \mid L\ HH\ M \mid 1.52$ (solid curve), $2.08 \mid L \mid 1$ (dotted curve labeled r_A), and $2.08 \mid M \mid 1.52$ (dotted curve labeled R_B), with $n_L = 1.38$, $n_H = 2.08$, and $n_M = 1.62$.

between the substrate and the surrounding medium index—regardless whether a material with this index exists in nature or not.

2. All layers have to be equally thick.

It is difficult to assess how much of a restriction the equal thickness

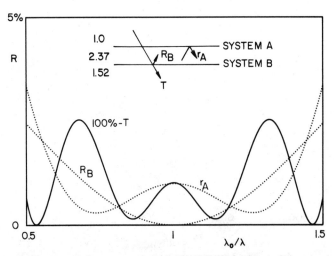

Figure 4.22.   Reflectance of $1 \mid L\ M1\ HH\ M2 \mid 1.52$ (solid curve), $2.37 \mid M1\ L \mid 1$ (dotted curve labeled r_A), and $2.37 \mid M2 \mid 1.52$ (dotted curve labeled R_B), with $n_L = 1.4$, $n_{M1} = 1.95$, and $n_{M2} = 1.90$, and $n_H = 2.37$.

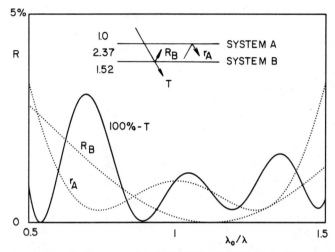

Figure 4.23. Reflectance of 1 | L M1 HH 0.9M2 | 1.52 (solid curve), 2.37 | M1 L | 1 (dotted curve labeled r_A), and 2.37 | 0.9M2 | 1.52 (dotted curve labeled R_B) with $n_L = 1.4$, $n_{M1} = 1.95$, $n_{M2} = 1.90$, and $n_H = 2.37$.

postulate (2) is. When all the layers are equally thick it follows from the symmetry of the trigonometric functions that

$$R(\lambda_0/\lambda = 1+a) = R(\lambda_0/\lambda = 1-a) \qquad (4.10)$$

This means in words: If the reflectance is low at a certain position,

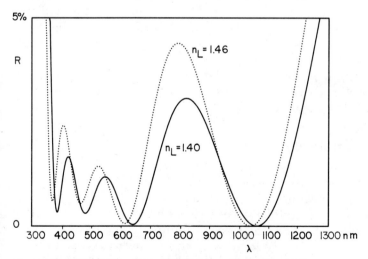

Figure 4.24. Reflectance of two dual-band antireflection coatings as proposed by Szafranek and Lubezky[7]: 1 | L M1 HH 0.9M2 | 1.52 with $n_L = 1.4$, $n_{M1} = 1.95$, $n_{M2} = 1.9$, $n_H = 2.37$, and $\lambda_0 = 570$ nm (solid curve) and $n_L = 1.46$, $n_{M1} = 2.05$, $n_{M2} = 1.93$, $n_H = 2.45$, and $\lambda_0 = 550$ nm (dotted curve).

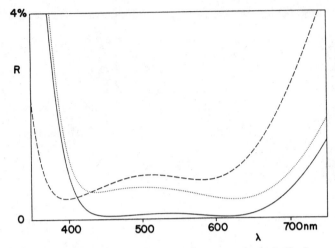

Figure 4.25. The reflectance at an angle averaged over both planes of polarization for the design 1.0 | L 2.11H 0.331L 0.225H | 1.52 at $\alpha = 0°$, $\beta = 0°$ (solid curve), 45°, 0° (dashed curve), and 45°, 45° (dotted curve). The first angle is the incidence angle and the second the match angle (Eq. 2.58). $n_L = 1.38$, $n_H = 2.08$, and $\lambda_0 = 520$ nm.

automatically it also is low at the position symmetrical to $\lambda_0/\lambda = 1$. This is quite an advantage if one wants to achieve low reflectance over a maximally wide region.

The first problem turns out to be a very severe restriction, especially when the surrounding medium is air. If for the first layer an index of less than 1.38 (MgF_2) is specified it cannot be realized with a hard and durable coating material in practice and the design has to be discarded. The limit of 1.38 holds true for the visual spectrum. It is higher for other wavelength regions. For the second part of this problem, i.e., availability of any index value between the first layer and the substrate, single-layer synthesis (Sec. 3.1.4) provides a solution.

4.5.1. Step-up antireflection coatings

Direct application of Chebyshev synthesis leads to designs where the refractive indices of the layers monotonously increase in steps from the surrounding medium to the substrate. We call them step-up coatings. Due to the nonavailability of coating materials with refractive indices smaller than 1.38, their use is limited to applications where the substrate index is very large (e.g., in the infrared). Examples of three designs with equal bandwidth but different numbers of layers are given in Fig. 4.10.The values were taken directly from Tables 3.5 to 3.8 (see also Baumeister[5a]).

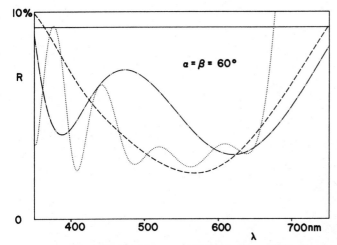

Figure 4.26. The reflectance at an angle averaged over both planes of polarization for the designs 1.0 | 1.105L 1.7HH | 1.52 (dash-dotted curve), 1.0 | 1.04L 1.52H 0.229L 0.214H | 1.52 (dashed curve), and 1.0 | 0.864L 1.69H 1.69L 1.69H 1.69L 1.92H | 1.52 (dotted curve). The reflectance of the uncoated glass surface is shown as a solid line. The match and incidence angles are 60° (λ_0 = 500 nm). n_L = 1.38 and n_H = 2.06.

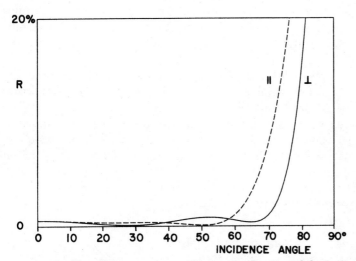

Figure 4.27. The reflectance at λ = 620 nm as a function of the incidence angle of the design 1.0 | 0.95L 3.91M3 0.796L 1.11M1 1.46H 1.13M2 | 1.55 with n_L = 1.33, n_{M1} = 1.52, n_{M2} = 1.74, n_{M3} = 2.10, and n_H = 2.3. (*Dobrowolski.*[10])

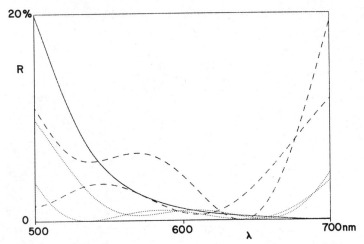

Figure 4.28. Reflectance as a function of wavelength at selected angles of the design in Fig. 4.25 ($\lambda_0 = 628$ nm). $\alpha = 0°$ (solid curve), $\alpha = 45°$ (dotted curves), and $\alpha = 60°$ (dashed curves). $\beta = 0°$ in all cases.

4.5.2. Step-Up–Step-Down Antireflection Coatings

Equal ripple designs for low-index substrates using coating materials with $n \geq 1.38$ are possible in a compound fashion with the introduction of an imagined, in reality not existing, so-called "dummy" medium. We call them step-up–step-down designs. Two designs from Tables 3.4 to 3.7 are combined, one to match from the medium (n_0) to the dummy medium (n_D) and another from the dummy medium (n_D) to the substrate (n_s). The split filter formula (Eq. 2.55) assures that $T = 1$ for r_A and $R_B = 0$.

Figure 4.11 gives an example of a wideband antireflection coating with $\lambda_2/\lambda_1 = 3$ on low-index glass ($n_s = 1.52$). For the *step-up* portion we determined, from Table 3.4, $n_D = n_S/n_0 = 2.61$ to be the lowest *substrate* index which yields the minimally allowable layer index of 1.38. For n_2 we then obtained $n_2 = 2.61/1.38 = 1.89$. For the *step-down* portion we entered Table 3.4 with $n_S/n_0 = 2.61/1.52$ and determined $n_3/1.52 = 1.42$ and $n_4/1.52 = 1.20$ or $n_3 = 2.18$ and $n_4 = 1.82$.

Figure 4.12 shows the reflectance of the design of Fig. 4.11 as a function of wavelength with $\lambda_0 = 600$ nm.

4.6. Antireflection Coatings and Imaged Equivalent Layers

In Sec. 3.1.6 we established that the reflectance of a nonsymmetric sequence of layers vanishes as long as the imaged equivalent indices

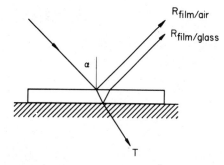

Figure 4.29. Light rays transmitted and reflected by a single
film when zigzag reflections are
neglected.

match the refractive indices of the surrounding medium and the substrate. We can use this theorem to design antireflection coatings.

As an example let us look at Fig. 3.1 for $n_L = 1.45$, $n_H = 2.35$, and
λ_0/λ from 1.5 to 2.5, $N(L/2 \text{ H } L/2) \approx 1$ and $N(H/2 \text{ L } H/2) \approx 3$. The two-
layer coating L/2 H/2 should consequently be an antireflection coating
between $n_0 = 1$ and $n_s = 3$. Figure 4.13 gives the verification. The
theorem can also be used to broaden existing designs. In Fig. 4.14 we
show the imaged equivalent indices of the $\lambda/4$–$\lambda/2$–$\lambda/4$ coating of Fig.
4.7. $N_{+\text{left image}}$ matches, except for two poles, $n_0 = 1$ over a wider λ_0/λ
region than $N_{+\text{right image}}$ matches $n_s = 1.52$. By inserting an LL layer
with $n_L = 1.38$ between the coating and the substrate a broader low-
reflectance region can be generated. The LL layer acts as an absentee

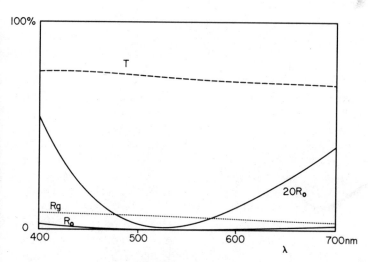

Figure 4.30. Theoretical reflectance and transmittance of a 75-nm-thick
film of index 1.46 over a 4-nm-thick nickel film on a glass substrate
($n = 1.52$). The thicknesses given are physical thicknesses. The optical
constants of nickel are from Table 5.3.

layer ($4\lambda/4$) at $\lambda_0/\lambda = 2$, as a $3\lambda/4$ layer at $\lambda_0/\lambda = 1.3$, and as a $5\lambda/4$ layer at $\lambda_0/\lambda = 2.7$. Figure 4.15 gives the result.

We note that the described procedure changed the step-up–step-down design into a step-up–step-down–step-up design—a trend which can also be observed with the exhaustive search designs of the next section.

4.7. Antireflection Coatings by Exhaustive Search

The number of layers in antireflection coatings is low enough so that systematic computer searches for the optimum coating are possible. The complexity can be reduced by making the optical thicknesses of all the layers equal (see discussion of symmetry, Eq. 4.10). The refractive indices can assume any value between two limits. Those refractive indices which do not exist in practice are later synthesized using Ohmer's formulas (Eqs. 3.15 and 3.16).

The calculations start at $\lambda_0/\lambda = 1$. If the reflectance is below a specified value the position $(\lambda_0/\lambda) + \Delta(\lambda_0/\lambda)$ is selected. If the reflectance is again below the specified value the procedure is repeated for the positions $(\lambda_0/\lambda) + 2\Delta(\lambda_0/\lambda)$, $(\lambda_0/\lambda) + 3\Delta(\lambda_0/\lambda)$, ... until either a specified bandwidth is reached (the design is then a solution) or the reflectance exceeds the specified value (the design is discarded). Figure 4.16 gives a six-layer design and its synthesization, Fig. 4.17 a seven-layer design (Thelen[6]), and Fig. 4.18 a nine-layer design.

4.8. Antireflection Coatings Using Higher Refractive Indices

Most antireflection coatings in the visible range use MgF_2 as a low-index material (see introduction to Chap. 4). But MgF_2 forms a hard and durable film only when it is deposited on a substrate at elevated temperature. Also, it cannot be prepared by cathode sputtering. As a consequence, applications to plastic substrates or large areas are not possible. A good coating material which does not have a refractive index quite as low as MgF_2 is SiO_2. It can be deposited at lower temperatures and also by sputtering. Its refractive index is 1.45.

Compared to antireflection coatings with $n_L = 1.38$ and $n_H = 2.1$ (Sec. 4.4) the same bandwidth can generally be achieved when we increase n_H to 2.35. In Fig. 4.19 we compare one of the designs of Fig. 4.7 ($n_L = 1.38$ and $n_H = 2.08 \approx 2.1$, solid curve) with a similar design using $n = 1.46$ and $n = 2.35$ (dashed curve). If higher residual reflectances can be tolerated and a greater bandwidth is desired a coating with $n = 1.46$ and 2.1 (dotted curve) can be used. In Fig. 4.20

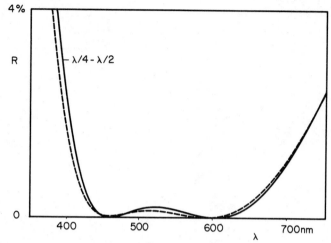

Figure 4.31. Reflectance of the two designs 1 | L HH | 1.74 (solid curve) and 1 | L 2.108H 0.188L 0.175L | 1.74 (dashed curve) with $n_L = 1.38$ and $n_H = 2.08$ ($\lambda_0 = 520$ nm).

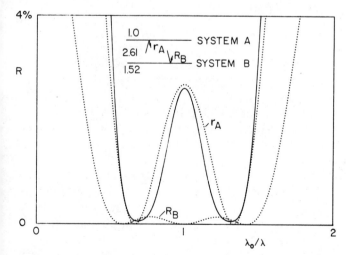

Figure 4.32. Reflectance of the design 1.0 | L M2 H M3 M1 | 1.52 with $n_L = 1.38$, $n_{M1} = 1.69$, $n_{M2} = 1.89$, $n_{M3} = 1.99$, and $n_H = 2.34$ (solid curve). Also shown as dotted curves are the reflectances 2.61 | M2 L | 1 (labelled r_A) and 2.61 | H M3 M1 | 1.52 (labelled R_B).

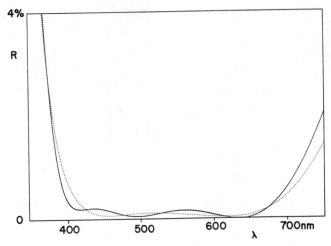

Figure 4.33. Reflectance of two antireflection coatings for high-index glass substrate (λ_0 = 520 nm): 1.0 | L 2.108H 0.331L 0.225H M1 | 1.75 (solid curve) and 1.0 | L 2.108H 0.331L 0.225H M2 M3 | 1.75 (dotted curve) with n_L = 1.38, n_H = 2.08, n_{M1} = 1.63, n_{M2} = 1.58, and n_{M3} = 1.69.

we show two designs with n_L = 1.45 based on the design method of Fig. 4.15.

4.9. Dual-Band Antireflection Coatings

Let us assume a coating to be low reflecting around λ_0 and constructed entirely of quarter-wave-thick layers. Due to the periodicity of the trigonometric functions of Eqs. 2.11 and 2.34 it must also be low reflecting around $\lambda_0/3$. It consequently is easy to design dual-band antireflection coatings when $\lambda_2 = 3\lambda_1$.

It is much more difficult to design a dual-band antireflection coating when $\lambda_2 = 2\lambda_1$. One possible solution would be a very wide antireflection coating as we discussed in Secs. 4.5 to 4.7. But this would be a rather extravagant approach.

Szafranek and Lubezky[7] proposed another way based on the method of effective interfaces (Sec. 3.4). The application of this method to the design of antireflection coating was extensively discussed before by Musset and Thelen.[3] Figure 4.21 shows how it is applied to the basic $\lambda/4-\lambda/2-\lambda/4$ coating of Sec. 4.3.1 (see also Thelen[8]).

In Fig. 4.22 a $\lambda/4-\lambda/4-\lambda/2-\lambda/4$ coating is analyzed. For direct application the reflectance would be too high. But if the design is skewed by shifting the single layer of system B to shorter wavelengths, one of the two large peaks is increased and the other decreased (Fig. 4.23).

Two versions of the resulting design are shown in Fig. 4.24. We will see later, Prob. 12.4, how Szafranek and Lubezky[7] converted these designs into six-layer two-material designs.

4.10. Antireflection Coatings at Nonnormal Light Incidence

Antireflection coatings are often used at nonnormal light incidence or in collimated light beams where the incidence angle varies. In Fig. 4.25 we can see that at 45° a severe deterioration of the good low-reflectance characteristic of normal incidence occurs. By matching the design to 45°, using Eq. 2.58, the reflectance at 45° comes down but now the reflectance at 0° is poor.

In Fig. 4.26 we show the reflectance characteristics of two-, four-, and six-layer antireflection coating designs for the rather high incidence angle of 60° (Snavely, Lewin, and Small[9]).

It is very difficult to design an antireflection coating which provides low reflectance over a large angular range. In Fig. 4.27 we show a design by Dobrowolski.[10] It keeps the reflectance low in both planes of polarization from 0 to 60°. Unfortunately, the spectral region of low reflectance is very narrow (Fig. 4.28).

4.11. Antireflection Coatings Using Absorbing Layers

In an ideal single-layer antireflection coating (Eq. 4.6) the reflectance of the film/air interface matches the reflectance of the film/glass interface (Fig. 4.29). When $n_{film} > \sqrt{n_S}$ (which is the case in practice since the lowest available index is $n_{MgF_2} = 1.38$) this can be compensated with a thin metallic film at the film/glass interface. Figure 4.30 gives an example.

This is, of course, only a simple example of what can be accomplished when different reflectances from both sides and lower transmittance can be tolerated. For more advanced treatment see Macleod.[11]

4.12. Problems and Solutions

Problem 4.1

For a substrate index of $n_S = 1.74$, compare a $\lambda/4$–$\lambda/2$ coating (Fig. 4.4) with a four-layer two-material coating according to Table 4.1. Use the same two coating materials for the $\lambda/4$–$\lambda/2$ coating as for the coating according to Table 4.1.

Solution. Figure 4.31 on p. 105 shows that the improvement with the two additional layers is not very substantial.

Problem 4.2

In Sec. 4.5.2 we used Table 3.8 to design the step-up–step-down antireflection coating 1.0 | L M1 H M2 | 1.52 (Fig. 4.11) by assuming a dummy medium with refractive index $n_D = 2.61$ between M1 and H. Instead of using two layers to match from the dummy medium to glass now use three, maintaining the same bandwidth (Table 3.5, $\lambda_0/\lambda = 3$).

Solution. For a substrate $n_S/n_0 = 2.61/1.52 = 1.72$ we obtain, from Table 3.5, the refractive index sequence: 1.0, 1.115, 1.311, 1.54, 1.72, or, after scaling to match from 1.52 to 2.61, 1.52, 1.69, 1.99, 2.34, 2.61. Surprisingly, the resulting five-layer antireflection coating of Fig. 4.32 (p. 105) is worse than the original four-layer coating of Fig. 4.11 (p. 92). There is a compensation of residual reflectances in the four-layer case but not in the five-layer case.

Problem 4.3

One of the four-layer two-material designs of Fig. 4.9 was designed for a substrate index of 1.52. Adapt this design to a substrate index of 1.75 by inserting one and two matching layers between the coating and the substrate: $n = \sqrt{1.52 \times 1.75} = 1.63$ (Thelen[12]).

Solution. For the double-layer solution, we determine from Table 3.4 (Collin's formulas) for $\lambda_2/\lambda_1 = 2.0$: 1.0, 1.04, 1.106, 1.15, or scaled 1.52, 1.58, 1.68, 1.75. Figure 4.33 on p. 106 shows the respective reflectances.

5

High Reflectors and Neutral Beam Splitters

5.1. High Reflectors

High reflectors are coatings which maximize the reflectance of an optical surface in a specified wavelength region. The optical performance outside the specified region is generally of no interest.

Thin films of silver, aluminum, gold, or rhodium have high reflectance over wide wavelength regions but are soft and sensitive to environmental attack. They require a protective layer (Hass and Scott[1]). Additional dielectric layers can increase the reflectance above what can be accomplished with the metal film alone (Fig. 5.1) (Hass[2]).

Often, metal films are not acceptable because of their high absorption and sensitivity to oxidation. Dielectric stacks are an alternative but at the price of a much narrower high-reflectance region. Figure 5.2 gives the reflectance of quarter-wave stacks with 15, 17, and 19 layers. The indices alternate between 2.35 and 1.45. By increasing the thickness of the next to the last layer (second from the substrate) from a quarter wave to a half wave, the reflectance curve can be flattened.

The high-reflectance zone of dielectric stacks can be widened by combining stacks with different center wavelengths (Sec. 6.3).

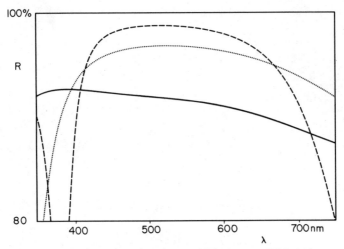

Figure 5.1. Reflectance of 1 | Al | 1.52 (solid curve), 1 | H L Al | 1.52 (dotted curve), and 1 | H L H L Al | 1.52 (dashed curve) with $n_H = 2.35$ and $n_L = 1.45$ ($\lambda_0 = 500$ nm). Al stands for a 50-nm-thick film of aluminum. Optical constants are from Table 5.3.

5.2. Neutral Beam Splitters

A neutral beam splitter is a device which splits an incident beam of light into two parts with a predetermined intensity ratio independent of wavelength λ over a desired wavelength range.

Thin metal films with or without nonabsorbing film enhancements make excellent neutral beam splitters. Figures 5.3 and 5.4 give two common designs by Pohlack.[3] In addition to their excellent neutrality, they show only small polarization at nonnormal incidence. On the negative side, though, high light loss, inability to handle a large thermal load, and limited oxidation stability precludes them from many applications.

Neutral beam splitters composed entirely of nonabsorbing films are an alternative. They can be made to be stable, to take large thermal load, and to suffer little from oxidation. But the bandwidth is much smaller, the number of layers much higher, and, if used at a high angle, the polarization effects are much more severe.

We also include in this section the design of partial reflectors as a neutral beam splitter with small bandwidth.

Neutral beam splitters at normal light incidence. At normal light incidence, the number of layers of neutral beam splitters is normally low enough to allow design by trial and error or exhaustive search methods, as was the case with antireflection coatings (Sec. 4.7).

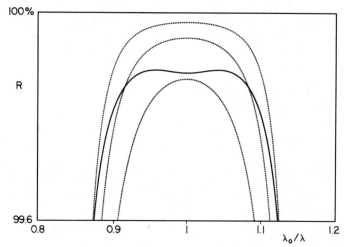

Figure 5.2. Reflectance of the designs 1 | (H L)^{p}H | 1.52 with $n_L = 1.45$ and $n_H = 2.35$, $p = 7, 8, 9$ (dotted curves, the higher is p, the higher is the peak reflectance). Solid curve is for the design 1 | (H L)^{9}LH | 1.52.

In Figure 5.5 we show designs with intermediate reflectance and in Fig. 5.6 designs with high reflectance. Some of the designs use standard coating materials (solid and dotted curves). The design of Fig. 5.5 with dashed reflectance curve uses refractive indices which still need synthesization. The design of Fig. 5.6 with dashed reflectance curve is a classical design by Baumeister and Stone.[4] It was the result of the first

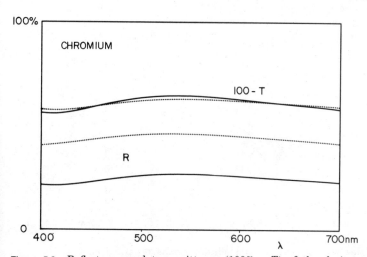

Figure 5.3. Reflectance and transmittance $(100\% - T)$ of the designs 1 | Cr | 1.52 (solid curves) and 1 | Cr H | 1.52 (dotted curves) with $n_H = 2.35$ and $\lambda_0 = 500$ nm. The chromium layers are 2.4 nm thick. The light incidence is normal. Optical constants are taken from Table 5.3.

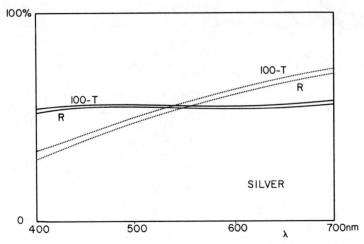

Figure 5.4. Reflectance and transmittance (100% − *T*) of the designs
1 | Ag | 1.52 (dotted curves) and 1 | Ag H | 1.52 (solid curves) with $n_H = 2.35$
and $\lambda_0 = 500$ nm. The silver layers are 17.5 nm thick. The light incidence
is normal. The optical constants were taken from Table 5.3.

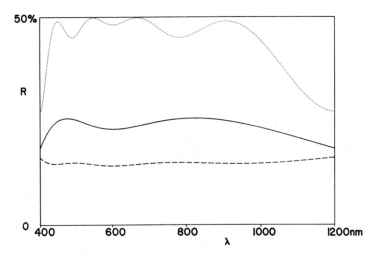

Figure 5.5. Reflectance of neutral beam splitter designs at normal light
incidence with intermediate reflectance (all $\lambda_0 = 600$ nm): 1 | H LL | 1.52
with $n_H = 2.08$ and $n_L = 1.38$ (solid curve), 1 | L HH LL H L H | 1.52 with
$n_L = 1.38$ and $n_H = 2.35$ (dotted curve), and 1 | H M1 M2 L | 1.52 with
$n_H = 2.17$, $n_{M1} = 1.98$, $n_{M2} = 1.75$, and $n_L = 1.59$ (dashed curve).

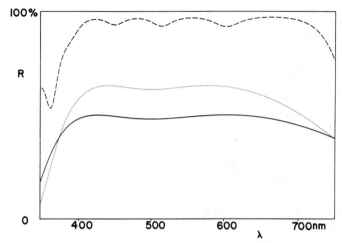

Figure 5.6. Reflectance of the neutral beam splitter designs at normal light incidence with high reflectance (all $\lambda_0 = 500$ nm): $1 \mid \text{L H L H LL} \mid 1.52$ with $n_L = 1.38$ and $n_H = 2.35$ (solid curve), $1 \mid \text{H L H M1 L M2} \mid 1.52$ with $n_H = 2.35$, $n_L = 1.38$, $n_{M1} = 1.44$, and $n_{M2} = 1.62$ (dotted curve), and $1 \mid 0.828\text{H } 0.828\text{L } 0.828\text{H } 0.87\text{L } 0.928\text{H}$ $0.928\text{L } 1.042\text{H } 1.034\text{L } 1.252\text{H } 1.402\text{L } 1.152\text{H } 1.334\text{L } 1.382\text{L}$ $1.382\text{H} \mid 1.52$ with $n_H = 2.3$ and $n_L = 1.35$ (dashed curve).

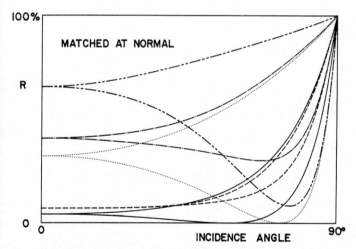

Figure 5.7. Reflectance as a function of the incidence angle of the combinations $1 \mid 1.52$ (solid curve), $1 \mid \text{L H} \mid 1.52$ (dashed curve), $1 \mid \text{H} \mid 1.52$ (dotted curve), $1 \mid \text{L H L H} \mid 1.52$ (dash-dotted curve), and $1 \mid \text{H L H} \mid 1.52$ (dash-dot-dotted curve). In all cases $n_L = 1.45$, $n_H = 2.35$, and $\lambda = \lambda_0$.

reported application of computer optimization techniques to optical interference coatings design.

5.3. Neutral Beam Splitters at Angle

Figure 5.7 gives the reflectance of some simple high-low combinations as a function of the incidence angle. Figure 5.8 gives the same combinations but now the match angle β (Eq. 2.58) is assumed equal to the incidence angle α (Lobsiger[5]). The spectral performance of the interesting 1 | L H L H | 1.52 design with its remarkable neutrality over both wavelength and angular regions is given in Fig. 5.9.

5.3.1. Designs based on nonpolarizing equivalent layers

Costich[6] introduced as a measure of polarization splitting the quantity

$$\Delta n = \frac{n_p}{n_s} = \frac{1}{\cos^2 \alpha} = \frac{1}{1 - (n_0 \sin \alpha_0 / n)^2} \tag{5.1}$$

and applied it to equivalent structures. By postulating $\Delta N = 1$ he could derive building blocks for nonpolarizing designs.

At $\lambda = 2\lambda_0$ the equivalent index of H/2 L H/2 is

$$N(2\lambda_0) = n_{\mathrm{H}} \sqrt{\frac{n_{\mathrm{H}}}{n_{\mathrm{L}}}} \tag{5.2}$$

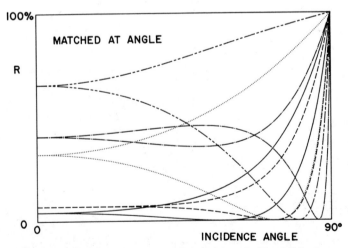

Figure 5.8. Reflectance of the same designs given in Fig. 5.7, but now the optical thicknesses are continuously adjusted so that for each incidence angle all layers remain $\lambda_0/4$ thick.

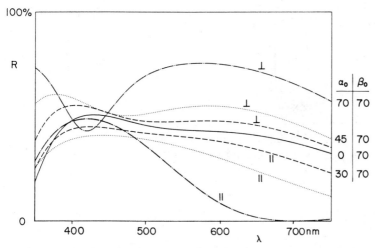

100%

R

α_0	β_0
70	70
45	70
0	70
30	70

Figure 5.9. Reflectance of the design 1 | L H L H | 1.52 with $n_L = 1.45$ and $n_H = 2.35$ ($\lambda_0 = 420$ nm, match angle $\beta = 70°$): $\alpha = 0°$ (solid curve), $\alpha = 30°$ (dashed curve), $\alpha = 45°$ (dotted curve), and $\alpha = 70°$ (dash-dotted curve). Higher values correspond to s-polarization.

and with (5.1)

$$\Delta N(2\lambda_0) = \Delta n_H \sqrt{\frac{\Delta n_H}{\Delta n_L}} \qquad (5.3)$$

Postulating $N(2\lambda_0) = 1$, we can derive from Eq. 5.3

$$\Delta n_L = \Delta n_H{}^3 \qquad (5.4)$$

Table 5.1 gives some values for an incidence angle of $\alpha = 45°$ and $n_0 = 1$. MgF$_2$/ZnS and ZnS/Ge satisfy criterion (Eq. 5.4) approximately.

TABLE 5.1 $\Delta n = 1/(1 - 1/2n^2)$

n	1.00	1.38	1.45	1.62	2.08	2.35	4.00
Δn	2.00	1.36	1.31	1.24	1.13	1.10	1.03

With the use of equivalent layers of the type H/2 L H /2 with n_H and n_L satisfying Eq. 5.4 we rid ourselves of the polarization splitting of the films. But the indices of the massive media are still split.

Now let us insert between a multilayer with characteristic matrix M_{11}, M_{12}, M_{21}, and M_{22} and its surrounding medium a quarter-wave-thick layer with refractive index n_{Q0}, and its substrate a quarter-wave-thick layer with refractive index n_{QS}. With Eq. 2.18, the right side of Eq. 2.33 takes the following form:

$$\begin{bmatrix} -in_{Q0} & \dfrac{-in_0}{n_{Q0}} \\[2mm] in_{Q0} & \dfrac{-in_0}{n_{Q0}} \end{bmatrix} \begin{bmatrix} M_{11} & iM_{12} \\[2mm] iM_{21} & M_{22} \end{bmatrix} \begin{bmatrix} \dfrac{in_S}{n_{QS}} & \dfrac{-in_S}{n_{QS}} \\[2mm] in_{QS} & in_{QS} \end{bmatrix} \begin{bmatrix} E^+ \\[2mm] E^- \end{bmatrix}$$

where n_{Q0} is the index of the quarter-wave layer on the surrounding-medium side and n_{QS} is the index on the substrate side. For the amplitude reflectance (Eq. 2.37), common factors of the matrix elements are of no significance. Eliminating in_0/n_{Q0} and in_S/n_{QS} and then setting

$$n_0' = \frac{n_{Q0}^2}{n_0} \quad \text{and} \quad n_S' = \frac{n_{QS}^2}{n_s} \tag{5.5}$$

we obtain

$$\vec{R}(n_0', M, n_S') = -\vec{R}(n_0, Q_0 M Q_S, n_S) \tag{5.6}$$

and $\qquad R(n_0', M, n_S') = R(n_0, Q_0 M Q_S, n_S)$

Thus, with Eqs. 5.1 to 5.5 we can avoid the polarization splitting of the massive media on both sides of the multilayer by postulating $\Delta n_0' = \Delta n_S' = 1$, or

$$\Delta n_{Q0} = \sqrt{\Delta n_0} \quad \text{and} \quad \Delta n_{QS} = \sqrt{\Delta n_S} \tag{5.7}$$

In Fig. 5.10 we show an infrared beam splitter design which Costich[6] gave in his pioneering paper to demonstrate his theory. Knittl and

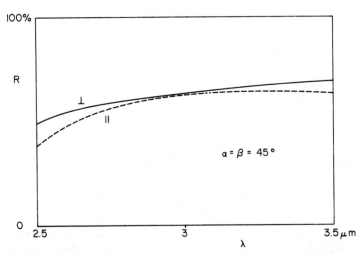

Figure 5.10. Reflectance of the design 1 | L 0.75H 1.5M2 0.75H M1 | 1.37 with $n_L = 1.37$, $n_{M1} = 1.7$, $n_{M2} = 2.35$, and $n_H = 4$ ($\lambda_0 = 3$ μm, $\alpha = \beta = 45°$).

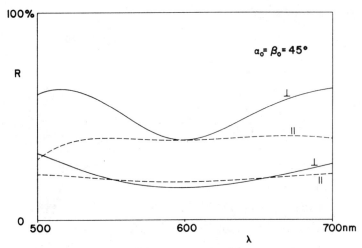

Figure 5.11. Reflectance of the designs 1 | M1 M4 M1 M4 M1 0.96H 1.08M2 0.96H M3 | 1.53 (upper curves) and 1 | L 0.96H 1.08M2 0.96H M3 | 1.53 (lower curves) with $n_L = 1.38$, $n_{M1} = 1.7$, $n_{M2} = 1.87$, $n_{M3} = 2.09$, $n_{M4} = 2.18$, and $n_H = 2.6$ ($\lambda_0 = 600$ nm, $\alpha = \beta = 45°$).

Houserková[7] extended Costich's theory. Figure 5.11 shows two of their designs.

5.3.2. All quarter-wave neutral beam splitters

In an early paper on the characteristics of quarter-wave stacks at nonnormal incidence, Baumeister[8] noted that the transmission and the degree of polarization can be expressed in terms of two variables. One variable is a function of the indices of refraction of the layers, the incident medium, and substrate; the other variable depends on the angle of incidence. The author (Thelen[9]) extended the theory to more than two materials.

The characteristic matrix of a quarter-wave stack composed of an odd number of layers with refractive indices n_1, n_2, n_2, . . . is

$$\mathbf{M}_{odd} = \begin{bmatrix} 0 & \dfrac{\pm i n_2 n_4 n_6 \ldots}{n_1 n_3\, n_5 \ldots} \\[2ex] \dfrac{\pm i n_1 n_3 n_5 \ldots}{n_2 n_4 n_6 \ldots} & 0 \end{bmatrix} \qquad (5.8)$$

For an even number of layers, the characteristic matrix is

$$\mathbf{M}_{even} = \begin{bmatrix} \dfrac{\pm n_2 n_4 n_6 \ldots}{n_1 n_3\, n_5 \ldots} & 0 \\[2ex] 0 & \dfrac{\pm n_1 n_3 n\,_5 \ldots}{n_2 n_4 n_6 \ldots} \end{bmatrix} \qquad (5.9)$$

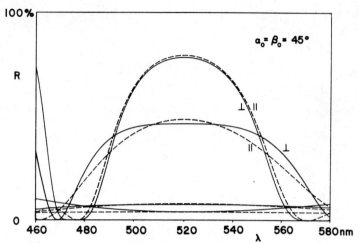

Figure 5.12. Reflectance of the designs 1.52 | L M H M H M L M L M H M H M H M L M L M H M H M L | 1.52 (R_{max} = 80%), n_L = 1.38, n_M = 1.66, and n_H = 2.35; 1.52 | L M H M H M L M L M H M H M L M | 1.52 (R_{max} = 50%), n_L = 1.38, n_M = 1.62, and n_H = 2.35; 1.7 | M L M H M L M | 1.7 (R_{max} = 8%), n_L = 1.35, n_M = 1.52, and n_H = 1.73; 1.52 | L M H M | 1.52 (R_{max} = 4%), n_L = 1.38, n_M = 1.61, and n_H = 2.3. λ_0 = 520 nm, α = β = 45°.

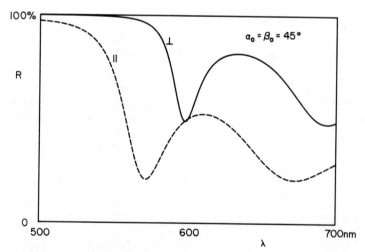

Figure 5.13. Reflectance of the design 1 | 0.778H L H 1.011L 0.333H L 0.622H L 0.333H 1.55L 0.311H L 0.622H L 0.311H | 1.52 with n_L = 1.38 and n_H = 2.35 (λ_0 = 600 nm, α = β = 45°).

With Eq. 2.40, the transmittance can be expressed in the following way:

$$T = \frac{4}{2 + X^2 + X^{-2}} \tag{5.10}$$

where

$$X_{\text{odd}} = \frac{\sqrt{n_0 n_S} \; n_2 n_4 n_6 \ldots}{n_1 n_3 n_5 \ldots} \tag{5.11}$$

$$X_{\text{even}} = \frac{\sqrt{n_0 / n_S} \; n_2 n_4 n_6 \ldots}{n_1 n_3 n_5 \ldots} \tag{5.12}$$

$T_p = T_s$ when $X_p = X_s$. With Costich's[4] terminology (Eq. 5.1), we obtain as conditions for a quarter-wave stack to be nonpolarizing at the quarter-wave point

$$\sqrt{\Delta n_0 \Delta n_s} \; \Delta n_2 \Delta n_4 \Delta n_6 \ldots$$
$$= \Delta n_1 \Delta n_3 \Delta n_5 \ldots \qquad \text{even number of layers} \tag{5.13}$$

and

$$\sqrt{\frac{\Delta n_0}{\Delta n_S}} \; \Delta n_2 \Delta n_4 \; \Delta n_6 \ldots$$
$$= \Delta n_1 \Delta n_3 \Delta n_5 \ldots \qquad \text{odd number of layers} \tag{5.14}$$

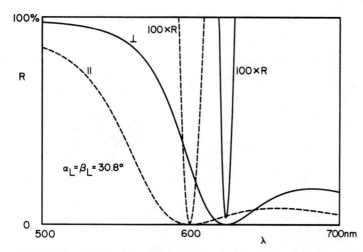

Figure 5.14. Reflectance of one part of the design of Fig. 5.13: 1.38 | 0.333H L 0.622H L 0.333H 1.55L 0.311H L 0.622H L 0.311H | 1.52 with $\alpha_L = \beta_L = 30.8°$. Other data are the same as in Fig. 5.13.

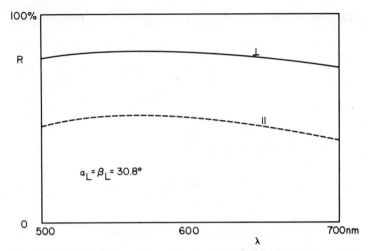

Figure 5.15. Reflectance of the other part of the design of Fig. 5.13:
1.38 | H L 0.778H | 1 with $\alpha_L = \beta_L = 30.8°$. Other data are the same as
in Fig. 5.13.

Figure 5.12 gives four examples (Thelen[9]). The theory presented here
was extended by Sterke, van der Laan, and Frankena.[10]

5.3.3. Partial reflectors using buffer layers

As discussed before (Sec. 3.5), a buffer layer has no effect on the overall
reflectance of a multilayer whatever its thickness (Mouchart[11] and
Knittl[12]). At angle, a buffer layer can only be effective in one plane or
polarization. In the other plane of polarization the reflectance will

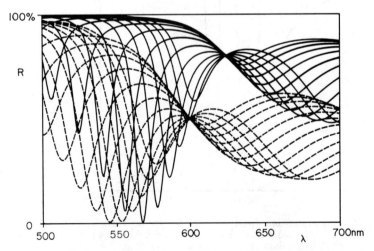

Figure 5.16. Reflectance of the design of Fig. 5.13. The fourth layer (1.011L)
varies now from 0 to 110 percent.

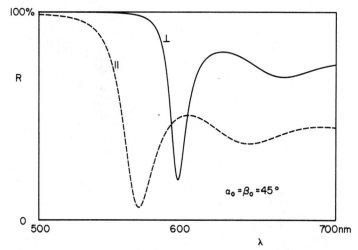

Figure 5.17. Reflectance of the design $1 \mid$ H L H L 0.793(H/2 L H/2)$^6 \mid 1.52$ with $n_H = 2.35$ and $n_L = 1.45$ ($\alpha_0 = \beta_0 = 45°$, $\lambda_0 = 600$ nm).

change when we change the thickness of the buffer layer. We consequently have a method to systematically change the reflectance in one plane of polarization while keeping it constant in the other (Knittl[12]).

Knittl's design method of a nonpolarizing partial reflector consists of three steps:

1. Select a design which has zero reflectance in one plane of polarization but a sufficiently high reflectance in the other, as seen from a suitable spacer material.

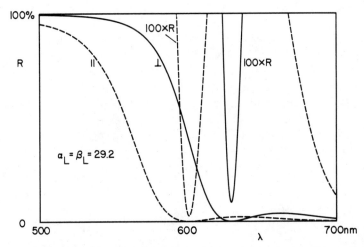

Figure 5.18. Reflectance of the edge polarizer design used in Fig. 5.17: $1.45 \mid 0.793$(H/2 L H/2)$^6 \mid 1.52$ with $\alpha_L = \beta_L = 29.2°$. Other data are the same as in Fig. 5.17.

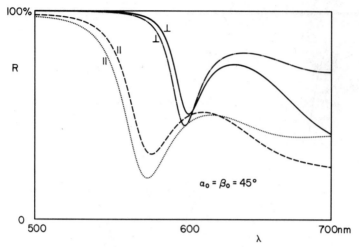

Figure 5.19. Reflectance of the designs $1 \mid$ H L H 1.133L 0.787(H/2 L H/2)$^5 \mid$ 1.52 (dash-dotted and dotted curves) and $1 \mid$ 0.35H L H L H 1.217L 0.787(H/2 L H/2)$^5 \mid$ 1.52 (solid and dashed curves) with $n_L = 1.45$ and $n_H = 2.35$ ($\lambda_0 = 600$ nm, $\alpha_0 = \beta_0 = 45°$).

2. Select another design which has the desired partial reflectance in the plane where the design selected in the first step has zero reflectance, as seen from the spacer material selected under step 1.

3. Adjust the spacer thickness until the reflectance in the plane with variable reflectance matches the other.

In Fig. 5.13 on p. 118, one of Knittl's[12] design is shown. 1.38 $\mid$ 0.333H L 0.622H L 0.333H 1.55L 0.311H L 0.622H L 0.311H $\mid$ 1.52 has zero reflectance in the p-plane (step 1, Fig. 5.14). 1.38 $\mid$ H L 0.778H $\mid$ 1 has 50 percent reflectance in the p-plane (step 2, Fig. 5.15). Adjustment of the spacer to 1.011 (step 3) produces the final result. In Fig. 5.16 the spacer of the design of Fig. 5.13 is varied from 0 to 110 percent.

An interesting feature of this design (Figs. 5.13 to 5.16) is the fact that at a slightly higher wavelength (620 nm) the buffer layer of the p-plane acts also as a buffer layer of the s-plane (Figs. 5.14 and 5.16).

In Fig. 5.17 another one of a Knittl's[12] designs is shown. Instead of a low index of 1.38 we use 1.45. The zero reflectance is now accomplished with an *edge polarizer* design (Sec. 9.1.2) (Fig. 5.18). Figure 5.19 gives optimized variations of the design of Fig. 5.17.

5.4. Problems and Solution

Problem 5.1

In Fig. 5.1 equally thick H- and L-layers were used. Take the phase shift upon reflectance on a metal layer into account and calculate how

TABLE 5.2 Reflectance of Two High-Reflector Designs

λ	Al │ 0.858L H │ 1.0	Al │ L H │ 1.0
542 nm	97.097%	96.922%
544	97.098	96.935
546	97.099	96.947
548	97.099	96.959
550	97.099	96.969
552	97.098	96.979
554	97.097	96.989
584	97.036	97.065
586	97.029	97.067
588	97.022	97.068
590	97.014	97.068
592	97.006	97.068
594	96.998	97.068
596	96.990	97.068
598	96.981	97.067
600	96.972	97.065

much the thickness of the dielectric layer next to the metal layer has to be decreased (increased) for optimum performance.

Solution. Let us assume the metal layer $(n_m - ik_m)$ to be infinitely thick. Then the reflectance into a medium n_d is given by

$$\vec{R} = \frac{n_d - n_m + ik_m}{n_d + n_m - ik_m} = \frac{[(n_d - n_m) + ik_m][(n_d + n_m) + ik_m]}{(n_d + n_m)^2 + k_m^2}$$

$$= \frac{n_d^2 - n_m^2 - k_m^2 + 2in_dk_m}{(n_d + n_m)^2 + k_m^2} = \sqrt{R}\, e^{i\Phi}$$

$$\tan \Phi = \frac{2n_dk_m}{n_d^2 - n_m^2 - k_m^2}$$

With $n_d = 1.38$ (magnesium fluoride), $n_m = 0.82$, and $k_m = 5.99$ (aluminum at 550 nm, Table 5.3): $\Phi = -22.51°$. From this we calculate for the phase thickness of the magnesium fluoride layer

$$\phi = \frac{180° - \Phi}{2} = \frac{154.49}{2} = 77.25°$$

and for the physical thickness

$$d = \frac{550 \text{ nm} \times 77.25°}{1.38 \times 360°} = 85.52 \text{ nm}$$

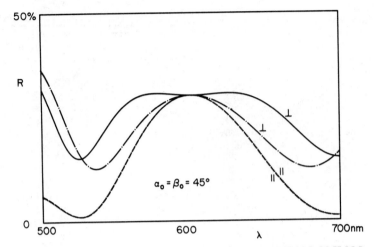

Figure 5.20. Reflectance of the designs 1.52 | L M H M H M L M H M L M H M | 1.52 (solid and dashed curves) and 1.52 | L M H M L M H M L M H M H M | 1.52 (dash-dotted and dotted curves) with $n_L = 1.45$, $n_M = 1.745$, and $n_H = 2.35$ ($\alpha_0 = \beta_0 = 45°$, $\lambda_0 = 600$ nm). The reflectances in the parallel planes appear to coincide.

instead of

$$d = \frac{550 \text{ nm} \times 90°}{1.38 \times 360°} = 99.64 \text{ nm}$$

In Table 5.2 (previous page) we show calculated reflectances near 550 nm and 600 nm for both cases. We observe that only the adjusted layer system peaks near the specified wavelength position. (The slight

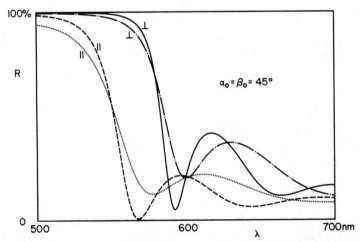

Figure 5.21. Reflectances of the two designs 1 | 0.27H L H 1.07L 0.793(H/2 L H/2)⁶ | 1.52 (solid and dashed curves) and 1 | 0.27H L H 1.53L 0.765(H/2 L H/2)⁵ | 1.52 (dash-dotted and dotted curves). Other data are the same as in Fig. 5.17.

discrepancy between 548 nm and 550 nm is due to the fact that we assumed an infinitely thick aluminum layer in our theory but only 100 nm in our calculation.)

Problem 5.2

Design an immersed beam splitter with $R = 30\%$, $n_0 = 1.52$, and $\alpha_0 = 45°$.

Solution.   With Eqs. 5.10 to 5.14 we find by trial and errror that three quarter waves with $n_1 = 1.45$ (odd positions), seven quarter waves with $n_2 = 1.745$ (even positions), and four quarter waves with $n = 2.35$ (odd positions) generate a reflectance of 30.2 percent. Figure 5.20 gives two versions of the design.

Problem 5.3

Modify the design of Fig. 5.17 for a reflectance of 20 percent instead of 50 percent.

Solution.   By reducing the thickness of the first layer to 0.27H the reflectance can be lowered to 20 percent (0.27H L H instead of H L H). The results are the solid and dashed curves of Fig. 5.21. By lowering the number of periods of the edge polarizer design from six to five a more stable design is generated (dash-dotted and dotted curves).

Problem 5.4

Establish the "current" values for the optical constants in the visual of silver, aluminum, chromium, and nickel from the literature.

Solution.

TABLE 5.3   Optical Constants for Ag, Al, Cr, and Ni

λ	Ag	Al	Cr	Ni
350 nm		3.25 − 3.91i		
400	0.075 − 1.93i	0.4 − 4.95i	3.8 − 4.2i	1.42 − 2.35i
450	0.055 − 2.42i	0.51 − 5i	4.5 − 4.4i	1.59 − 2.60i
492		0.64 − 5.5i		
500	0.05 − 2.87i		5.4 − 4.8i	1.74 − 2.96i
546		0.82 − 5.99i		
550	0.055 − 3.32i		5.9 − 5i	1.87 − 3.32i
578		0.93 − 6.33i		
600	0.06 − 3.75i		6.15 − 4.8i	1.96 − 3.62i
650	0.07 − 4.2i	1.3 − 7.11i	6.25 − 4.8i	2.05 − 3.82i
700	0.075 − 4.62i	1.55 − 7i	6.3 − 4.7i	2.15 − 3.99i
750		1.8 − 7.12i		

From G. Hass and L. Hadley. *Optical Properties of Metals, American Institute of Physics Handbook,* 3d ed., D. E. Gray, ed. McGraw-Hill, New York, 1982 reissue.

Edge Filters

Edge filters split polychromatic light into two parts according to wavelength: one part contains the light with all the longer wavelengths and the other part the light with all the shorter wavelengths. One part is transmitted and the other reflected and/or absorbed. Depending on whether the transmitted part contains the higher or lower wavelengths we call the edge filter a long or short wave pass.

Optical interference coatings with alternating high and low indices of refraction make ideal edge filters since their transmittance characteristics consist of a series of passbands and stopbands (Chap. 3). The art of designing edge filters consists in clearing the passband from secondary reflectance bands, making the transition from passband to stopband as sharp as possible, and extending the passband and stopband beyond their usual periodic limits. For complete rejection of unwanted light, absorbing substrates and absorption filters in series are often used.

Advanced edge filter design often requires the incorporation of techniques discussed in later chapters (Chap. 7, Minus Filters, and Chap. 8, Wideband Pass Filters). The reader is referred to these chapters for short-wave-pass edge filters which extended passband and/or rejection regions.

6.1. Edge Filter Design with Equivalent Layers

In order to achieve, through constructive interference, the rejection level that is normally required for edge filters, many layers are necessary. Since each layer has two parameters (refractive index and optical thickness), the design of an edge filter with, for example, 31 layers becomes mathematically a 62-parameter problem!

Equivalent layer theory (Sec. 3.1) reduces this problem to a three-layer problem:

1. Select an equivalent layer and repeat it until the steepness of transition from stopband to passband and the rejection level in the stopband fit the specification.

2. Design an antireflection coating to match the equivalent layer to the incident medium.

3. Design an antireflection coating to match the equivalent layer to the substrate.

Sometimes, the equivalent index of the equivalent layer of the core stack is close enough to the index of the incident medium and/or the substrate that one or both of these antireflection coatings can be omitted.

6.1.1. Stopband characteristics of equivalent layers

In Prob. 2.4 we derived formulas for the transmittance of quarter-wave stacks at the quarter-wave point. They can be used for rough estimates of the rejection level at the center of the stopband.

The transmittance at the edges of a stopband can be estimated in the following way. Let us assume that the characteristic matrix of the equivalent layer is given by

$$\mathbf{M}_E = \begin{bmatrix} M_{11} & iM_{12} \\ iM_{21} & M_{11} \end{bmatrix}$$

Then

$$\Gamma = \arccos M_{11} \quad \text{and} \quad N = \sqrt{\frac{M_{21}}{M_{12}}} = \frac{\sin \Gamma}{M_{12}} = \frac{M_{21}}{\sin \Gamma} \quad (6.1)$$

At the edges of the stopbands

$$\cos \Gamma \to 1 \quad \sin \Gamma \to 0 \quad N \to 0 \quad \text{or} \quad \infty \quad (6.2)$$

If we repeat the equivalent layer p times we obtain for the characteristic matrix of the resulting multilayer

$$\mathbf{M} = \mathbf{M}_E^p = \begin{bmatrix} \cos p\Gamma & \dfrac{i \sin p\Gamma}{N} \\ iN \sin p\Gamma & \cos p\Gamma \end{bmatrix}$$

or

$$\mathbf{M} = \mathbf{M}_E^p = \begin{bmatrix} \cos p\Gamma & iM_{12}\dfrac{\sin p\Gamma}{\sin \Gamma} \\ iM_{21}\dfrac{\sin p\Gamma}{\sin \Gamma} & \cos p\Gamma \end{bmatrix}$$

Since $(\sin p\Gamma)/\sin \Gamma \to p$ for $\sin \Gamma \to 0$:

$$\mathbf{M}_{\text{stopband edges}} = \begin{bmatrix} 1 & ipM_{12} \\ ipM_{21} & 1 \end{bmatrix} \tag{6.3}$$

Inserting the matrix elements of Eq. 6.3 into Eq. 2.40 we obtain for the transmittance at the stopband edges

$$T_{\text{stopband edges}} = \frac{4n_0 n_s}{(n_0 + n_S)^2 + p^2(n_0 n_S M_{12} + M_{21})^2} \tag{6.4}$$

We note that the transmittance at the stopband edges decreases with $1/p^2$.

Equation 6.4 is further simplified by the fact that either M_{12} (if $N \to 0$) or M_{21} (if $N \to \infty$) equals zero (follows from Eqs. 6.2 and 2.23).

$$N \to 0: \qquad T_{\text{stopband edge}} = \frac{4n_0 n_s}{(n_0 + n_S)^2 + (pn_0 n_S M_{12})^2} \tag{6.5}$$

$$N \to \infty: \qquad T_{\text{stopband edge}} = \frac{4n_0 n_S}{(n_0 + n_S)^2 + (pM_{12})^2} \tag{6.6}$$

In Fig. 6.1 the equivalent index of an H/2 L H/2 structure is superimposed upon the reflectance of a 7- and 11-layer long wave pass constructed with it. The stopband is hatched. Equation 6.5 is valid for points A and B and Eq. 6.6 for points C and D.

6.1.2. Matching in the passband away from the edge

Away from the stopbands, equivalent indices normally have only moderate dispersion. Single- and two-layer antireflection coatings are generally used here to match the equivalent layer to the incident medium and the substrate.

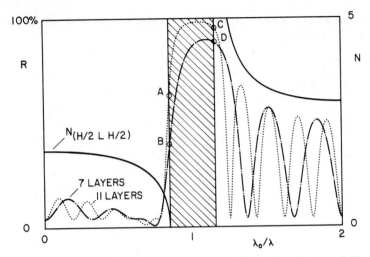

Figure 6.1. Equivalent index of the structure H/2 L H/2 with n_H = 2.35 and n_L = 1.45 superimposed upon the reflectance of 1 | (H/2 L H/2)^p | 1.52 with p = 3 (solid curve) and p = 5 (dotted curve). Stopband is hatched. A, B, C, and D mark the reflectances at the band edges.

As a demonstration, let us take the structure (L/2 H L/2)^9, centered at 450 nm and deposited on glass with refractive index 1.52, n_L = 1.45, and n_H = 2.35. Let us maximize the transmittance at 600 nm towards air (Fig. 6.2). As in Prob. 3.1, we determine for the equivalent index that N_{600} = 2.7. A proper matching layer towards air would be a layer,

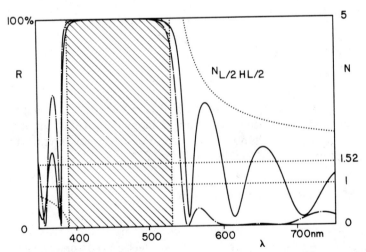

Figure 6.2. Reflectance of the long wave passes 1 | (L/2 H L/2)^9 | 1.52 and 1 | 1.33A (L/2 H L/2)^9 1.33B | 1.52 with n_A = 1.65, n_L = 1.45, n_H = 2.35, and n_B = 2.03 (λ_0 = 450 nm).

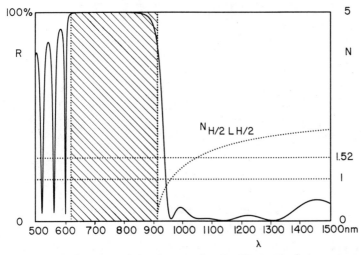

Figure 6.3. Reflectance of the design $1 \mid 1.5L1 \, (H/2 \, L2 \, H/2)^8 \mid 1.52$ with $n_{L1} = 1.38$, $n_{L2} = 1.85$, and $n_H = 3.4$ ($\lambda_0 = 740$ nm).

$\lambda/4$ thick at 600 nm (1.33×450 nm), having a refractive index $n_A = \sqrt{2.7 \times 1} = 1.65$. Towards glass with index 1.52, the matching layer would have the same thickness, but its refractive index would be $n_B = \sqrt{2.7 \times 1.52} = 2.03$.

Figure 6.3 gives a long-wave-pass design using silicon ($n = 3.4$) and SiO ($n = 1.85$) as base materials. As a bulk material, silicon has an absorption edge at 1.1 μm. As an evaporated layer, though, its ab-

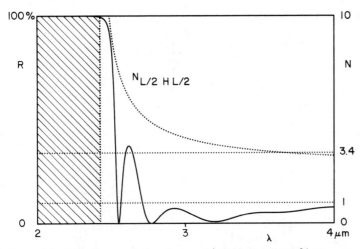

Figure 6.4. Reflectance of the design $1 \mid 1.55L \, (L/2 \, H \, L/2)^9 \mid 3.4$ with $n_L = 1.85$ and $n_H = 4.2$ ($\lambda_0 = 1.81$ μm).

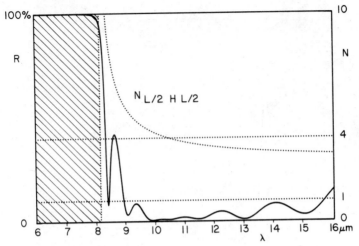

Figure 6.5. Reflectance of the design 1 | 1.58L 3.16H (L/2 H L/2)^{10} | 4 with $n_L = 2.2$ and $n_H = 4$ ($\lambda = 6.64$ μm).

sorption is negligible down to 0.9 μm. As a matching layer, MgF$_2$ ($n = 1.38$) is used. In Prob. 6.1 we will replace MgF$_2$ with a synthesized layer of silicon and SiO.

Figures 6.4 (SiO and Ge on silicon) and 6.5 (ZnS and Ge on Ge) give designs for the intermediate infrared region. Both of these designs have unacceptable secondary reflectance peaks near the edge. We will learn how to eliminate these peaks in the next section (Figs. 6.8 and 6.9).

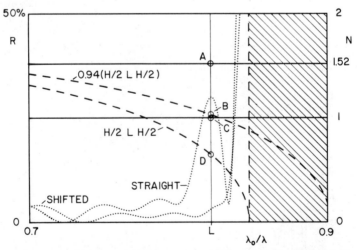

Figure 6.6. Equivalent index of the structures H/2 L H/2 and 0.94(H/2 L H/2) superimposed upon the reflectance of 1 | (H/2 L H/2)^{15} | 1.52 (labeled "straight") and 1 | (H/2 L H/2)^{12} 0.94(H/2 L H/2)^3 | 1.52 (labeled "shifted") with $n_H = 2.35$ and $n_L = 1.45$.

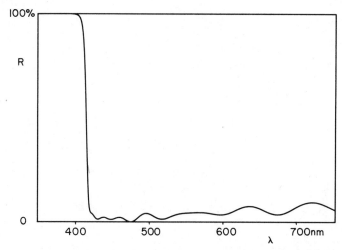

Figure 6.7. Reflectance of the long wave pass $1 \mid (H/2 \ L \ H/2)^{12} \ 0.94(H/2 \ L \ H/2)^3 \mid 1.52$ with $n_H = 2.35$ and $n_L = 1.45$ ($\lambda_0 = 350$ nm).

The interesting feature of the design of Figure 6.5 is the use of a two-layer antireflection coating.

6.1.3. Shifted equivalent layers

In Fig. 6.6 we show the reflectance of the design $1 \mid (H/2 \ L \ H/2)^{15} \mid 1.52$ (labeled "straight") with $n_H = 2.35$ and $n_L = 1.45$. The secondary re-

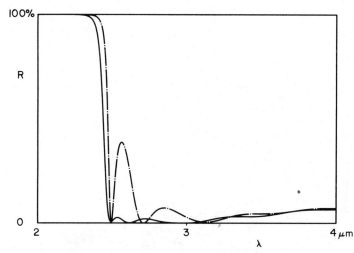

Figure 6.8. Reflectance of the design $1 \mid 1.55L \ 0.94(L/2 \ H \ L/2)^2 \ (L/2 \ H \ L/2)^5 \ 0.94 \ (L/2 \ H \ L/2)^2 \mid 3.4$ with $n_L = 1.85$ and $n_H = 4.2$ ($\lambda_0 = 1.81$ μm) (solid curve), compared to the design of Fig. 6.4 (dash-dotted curve).

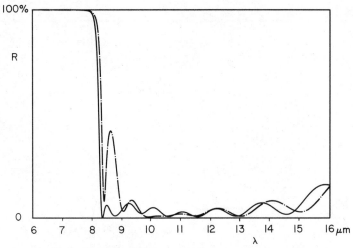

Figure 6.9. Reflectance of the design 1 | 1.58L 3.16H 0.9(L/2 H L/2)^2 (L/2 H L/2)^8 0.93 (L/2 H L/2)^2 | 4 with n_L = 2.2 and n_H = 4 (λ_0 = 6.64 μm)(solid curve), compared to the design of Fig. 6.5 (dash-dotted curve).

flectance peak before the edge has a reflectance of approximately 30 percent! Before, at a lower number of periods (p = 3 and 5 in Fig. 6.1 vs. p = 15 in Fig. 6.6), this problem did not exist.

The magnitude of the last secondary reflectance maximum before the edge is determined by the equivalent index at the location of the maximum. For small p this location is in the flat part of the $N(\lambda_0/\lambda)$-curve. For

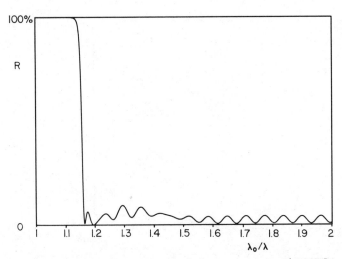

Figure 6.10. Reflectance of the visual short wave pass 1 | 1.065(L/2 H L/2)^2 (L/2 H L/2)^{11} 1.08 (L/2 H L/2)^2 | 1.52 with n_L = 1.45 and n_H = 2.35.

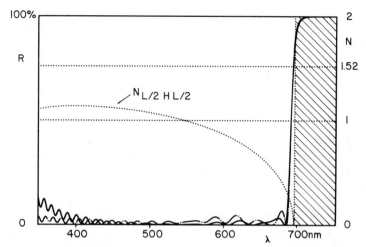

Figure 6.11. Reflectance of the visual short wave passes 1 | 1.065(L2/2 H L2/2)² (L2/2 H L2/2)¹¹ 1.065 (L2/2 H L2/2)² | 0.75L1 | 1.52 (dash-dotted curve) and   1 | 1.065(L2/2 H L2/2)² (L2/2 H L2/2)¹¹ 1.065   (L2/2 H L2/2)² 0.65(L1 2M1 M2) | 1.52 (solid curve) with n_{L1} = 1.38, n_{L2} =   1.45, n_{M1} = 2.1, n_{M2} = 1.63, and n_H = 2.35 (λ_0 = 800 nm).

large p it is in the steep portion of the curve. The easiest way to eliminate this peak is to use some of the equivalent layers to antireflect the rest. In Fig. 6.6 we see that the equivalent index of the shifted structure 0.94(H/2 L H/2) (point B) is approximately equal to the square root of the product of the equivalent index of the structure (H/2 L H/2) (point D) and the substrate index (point A). The shifted structure consequently fulfills the amplitude condition of antireflection (Eq. 4.6) on the substrate side. On the incident-medium side, we neglect the index mismatch.

In order to fulfill the phase condition of antireflection (second part of Eq. 4.6) we have to make sure that $\Gamma(\lambda_0/\lambda = L)$ is an odd multiple of $\lambda/4$. This is approximately the case when we repeat the structure three times. The final design is 1 | (H/2 L H/2)¹² | 0.94 (H/2 L H/2)³ | 1.52 as shown in Fig. 6.7. The same method can be applied to improve the near-edge performance of the designs of Figs. 6.4 and 6.5. The corresponding improvements are shown in Figs. 6.8 and 6.9.

Relative to long wave passes, the shift of short wave passes has to be reversed. Figure 6.10 gives the corresponding design to Fig. 6.7. In Fig. 6.11 two improved versions are shown: one with a single-layer antireflection coating and the other one with a triple layer. They look complex but they are easy to produce: just deposit the short wave pass on a glass substrate already antireflection coated!

In Figs. 6.6 to 6.10 we used the shifted equivalent layer in a single-layer antireflection coating mode—in spite of the fact that we had to

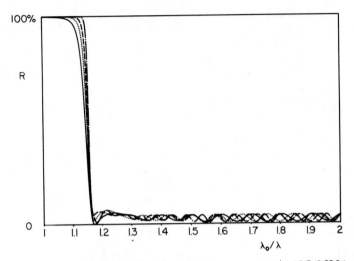

Figure 6.12. Reflectance of the visual short wave passes 1 | 1.12(L/2 H L/ 2) 1.06(L/2 H L/2) 1.03 (L/2 H L/2) 1.015(L/2 H L/2) (L/2 H L/2)^p 1.015(L/ 2 H L/2) 1.03(L/2 H L/2) 1.06(L/2 H L/2)   1.12(L/2 H L/2) | 1.52   with $n_L = 1.45$, $n_H = 2.35$, and $p = 1$ (solid curve), $p = 3$ (dotted curve), $p = 5$ (dashed curve), and $p = 10$ (dash-dotted curve).

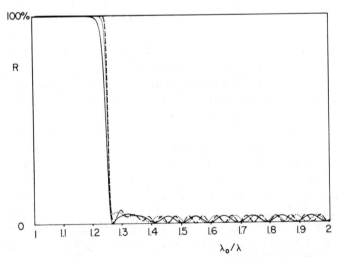

Figure 6.13. Reflectance of the infrared short wave passes 1 | 1.08(L/ 2 H L/2) 1.04(L/2 H L/2) 1.02   (L/2 H L/2) 1.01(L/2 H L/2) (L/2 H L/2)^p 1.01(L/2 H L/2) 1.02(L/2 H L/2) 1.04(L/2 H L/2)   1.08(L/2 H L/2) | 1.52 with $n_L = 1.90$, $n_H = 4.3$, and $p = 1$ (solid curve), $p = 3$ (dotted curve), $p = 5$ (dashed curve), and $p = 10$ (dash-dotted curve).

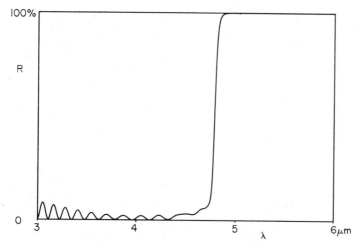

Figure 6.14.   Reflectance of the design of Fig. 6.13 on a different substrate
1 | 1.08(L/2 H L/2) 1.04(L/2 H L/2) 1.02   (L/2 H L/2) 1.01(L/2 H L/2) (L/2 H L/2)^5 1.01(L/2 H L/2) 1.02(L/2 H L/2) 1.04(L/2 H L/2) 1.08(L/2 H L/2) 0.667L | 2.7 with $n_L = 1.90$, $n_H = 4.3$ ($\lambda_0 = 6$ μm).

repeat it two to three times in order to generate the necessary multiple of 90° of the equivalent thickness. Now, instead of shifting the equivalent layers by the same amount we could stagger the shifts and generate a transition-type antireflection coating. In Figs. 6.12 and 6.13 we show two short-wave-pass designs, one for the visual region with indices 1.45 and 2.35, and one for the infrared region with indices 1.9 and 4.3. Both designs use four equivalent layers. The amounts of the relative shifts are not very critical as long as they are in the right direction (>1 in this configuration of equivalent and massive media indices). As seen from the massive media we reduced the step by 50 percent each time. The amount of the first step is a compromise between the steepness of the slope and the magnitude of the first ripple. Surprisingly, the match is so stable that the same outer system can be used for repetitions of the core system from 1 to 10. Figure 6.14 gives the design of Fig. 6.13 on a different substrate.

6.2.   Equal Ripple Edge Filter Design

Equal ripple prototype filters, as, for example, listed in Tables 3.11 and 3.12, are ideal edge filters. Their translation into practical designs, though, is difficult and normally requires many compromises. There are two major problems:

1. The theoretical designs are matched to specific substrate and incident-medium refractive indices. As a consequence, matching to the given substrate and medium indices is necessary.

TABLE 6.1 Translation of the 31-Layer Equal Ripple Design of Table 3.11 into Two-Material Designs†

		Matched at	
Layer number	Index	$\lambda_0/\lambda = 0.8$ $\lambda_0 = 350$ nm	$\lambda_0/\lambda = 1.2$ $\lambda_0 = 650$ nm
1	1.38	525	455
2	2.05	1050	960
3	1.45	82	205
4	2.35	174	216
5	1.45	213	474
6	2.35	81	100
7	1.45	193	449
8	2.35	214	265
9	1.45	209	469
10	2.35	52	65
11	1.45	190	443
12	2.35	256	316
13	1.45	202	458
14	2.35	30	36
15	1.45	186	437
16	2.35	290	359
17	1.45	194	447
18	2.35	16	20
19	1.45	182	432
20	2.35	316	391
21	1.45	186	437
22	2.35	8	9
23	1.45	179	427
24	2.35	333	412
25	1.45	181	430
26	2.35	3	4
27	1.45	176	424
28	2.35	343	425
29	1.45	178	426
30	2.35	1	1
31	1.45	175	423
32	2.35	349	432
layer 33 = layer 31, layer 34 = layer 30, ..., layer 61 = layer 3			
62	1.62	525	455
Substrate	1.52		

†Matching layers 1, 2, and 62 use additional standard visual coating materials.

2. The stepped refractive indices of the core design have to be synthesized with equivalent layers composed of available coating materials.

Both operations narrow the usable bandwidth considerably.

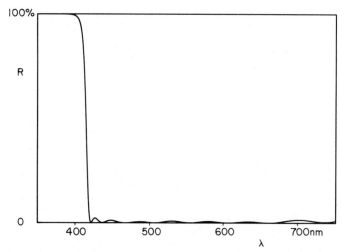

Figure 6.15 Reflectance of a long-wave-pass filter was specified under "Matched at $\lambda_0/\lambda = 0.8$" in Table 6.1 ($\lambda_0 = 350$ nm).

In Table 6.1 we show the translation of the 31-layer design of Table 3.11 to a long- and short-wave-pass filter for the visual region. Figures 6.15 and 6.16 give the corresponding reflectance curves.

In Fig. 6.17 we show the translation of the 17-layer infrared design of Table 3.12 to a long-wave-pass filter. Coating materials are Ge and SiO. Since the refractive index of SiO in the infrared very much depends on the speed of deposition, slow ($n = 1.6$), average ($n = 1.85$), and fast ($n = 1.95$) values can be used.

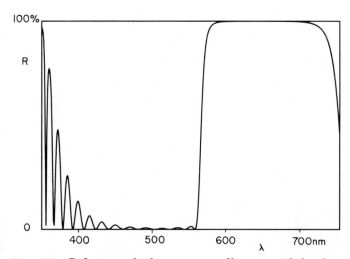

Figure 6.16. Reflectance of a short-wave-pass filter as specified under "Matched at $\lambda_0/\lambda = 1.2$" in Table 6.1 ($\lambda_0 = 650$ nm).

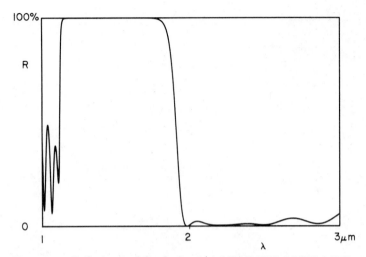

Figure 6.17. Reflectance of the design 1 | 1.64L1 0.185L 0.548H 0.557L 0.2H 0.471L 0.744H 0.543L 0.084H 0.485L 0.89H 0.521L 0.03H 0.48L (HL)⁴H 0.48L 0.03H 0.521L 0.89H 0.485L 0.084H 0.543L 0.744H 0.471L 0.2H 0.557L 0.548H 0.185L 1.86L2 | 1.45 with $n_{L1} = 1.6$, $n_L = 1.85$, $n_{L2} = 1.95$, and $n_H = 4.2$ ($\lambda_0 = 1.4$ μm).

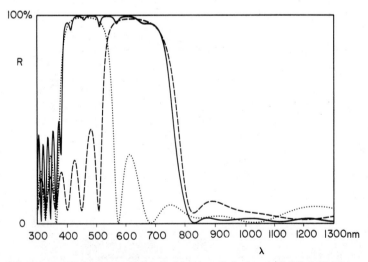

Figure 6.18. Reflectance of the "cold mirror" design 1 | 1.39L 0.714(L/ 2 H L/2)⁶ (L/2 H L/2)⁵ (L/2 H/2 L/2) | 1.52 (solid curve), 1.45 | 0.714 (L/ 2 H L/2)⁶ 1.39L | 1 (dotted curve), and 1.45 (L/2 H L/2)⁵ (L/2 H/2 L/ 2) | 1.52 (dashed curve) with $n_L = 1.45$, and $n_H = 2.35$ ($\lambda_0 = 630$ nm).

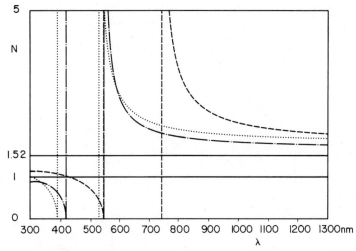

Figure 6.19. Equivalent indices of the configurations L/2 H L/2 ($\lambda_0 = 450$ nm, dotted curve), L/2 H L/2 ($\lambda_0 = 630$ nm, dashed curve), and L/2 H/2 L/2 ($\lambda_0 = 630$ nm, dash-dotted curve) with $n_L = 1.45$ and $n_H = 2.35$.

6.3. Edge Filters with Extended Rejection Region

Often, the stopband of a multilayer stack is not wide enough to provide rejection or high reflection over a broad enough wavelength region.

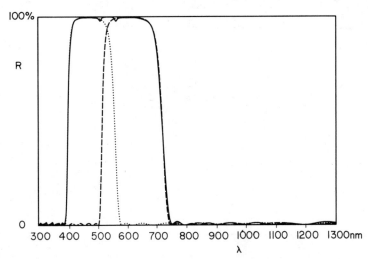

Figure 6.20. Reflectance of an equal ripple cold mirror composed of two of the 19-layer equal ripple stacks (called ERS) of Table 3.11: 1.671 | 0.775(ERS) 0.833M (ERS) | 1.671 (solid curve), 1.671 | 0.775(ERS) | 1.671 (dotted curve), and 1.671 | (ERS) | 1.671 (dashed curve) with $n_M = 1.671$ and $\lambda_0 = 600$ nm.

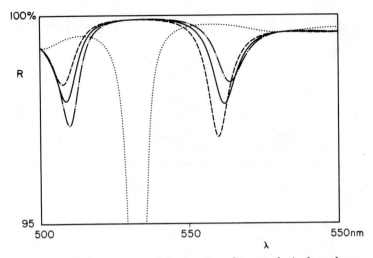

Figure 6.21. Reflectance around the junction of two stacks in dependence on the thickness of the spacer: $1.671 \mid 0.775(ERS) \, aM \, (ERS) \mid 1.671$ with $n_M = 1.571$, $\lambda_0 = 600$ nm, and $a = 0$ (dotted curve), $a = 0.750$ (dashed curve), $a = 0.833$ (solid curve), and $a = 0.917$ (dash-dotted curve).

The classical example is the so-called "cold mirror," a dichroic filter which reflects all the visual radiation and transmits as much of the infrared as the application for "cold" light requires. Generally, two stacks, centered at different wavelengths, are deposited on top of each other. Figure 6.18 gives an example. In Fig. 6.19 we show the equiv-

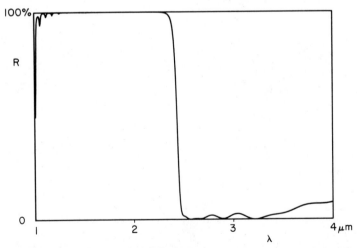

Figure 6.22. Reflectance of the design of Fig. 6.8 augmented with an additional quarter-wave stack: $1 \mid 1.55L \, 0.718(L/2 \, H \, L/2)^5 \, 0.94(L/2 \, H \, L/2)^2 \, (L/2 \, H \, L/2)^5 \, 0.94(L/2 \, H \, L/2)^2 \mid 3.4$ with $n_L = 1.85$, $n_H = 4.2$, and $\lambda_0 = 1.81$ μm.

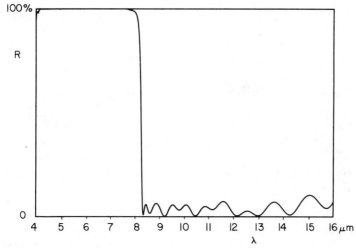

Figure 6.23. Reflectance of the design of Fig. 6.9 augmented with an additional quarter-wave stack: 1 | 1.58L 3.16H 0.726(L/2 H L/2)^8 0.9(L/2 H L/2)^2 (L/2 H L/2)^8 0.93(L/2 H L/2)^2 | 4 with n_L = 2.2, n_H = 4, and λ_0 = 6.64 μm.

alent indices of the structures used. The structure L/2 H/2 L/2 provides good matching between the stacks and the substrate.

Which stack should be on top? If the cold mirror is used as a front-surface reflector the stack with the thinner layers and/or the stack which reflects in the region of higher absorption (Chap. 12) should be the first as seen from the light incidence side. For a cold mirror this is definitely the blue reflecting stack.

The high-reflectance ripple is caused by two effects:

1. Interference between the main reflectance band of one of the stacks with secondary reflectance bands of the other stack
2. Interference at the wavelength position where the two stacks join

In order to separate these two effects let us go through the academic exercise of designing a cold mirror by combining two stacks which have virtually no secondary reflectance bands, for example, the 19-layer equal ripple designs of Table 3.11. We find (Fig. 6.20) that the rejection band ripple is eliminated except at two wavelength positions near the junction of the two stacks. This ripple can be controlled by optimizing the relative position of the two stack centers and the optical thickness of the spacer layer M (Fig. 6.21). For more detailed studies, split filter analysis (Sec. 2.9) is the proper tool.

In Figs. 6.22 and 6.23 the designs of Fig. 6.8 and 6.9 are augmented with an additional stack to broaden the rejection band. Note that in both cases the additional stack is put before (as seen from the light incidence side) the already existing stack.

6.4. Problems and Solutions

Problem 6.1

Replace the MgF$_2$-layer ($n = 1.38$) of the design of Fig. 6.3 with an equivalent layer composed of silicon ($n = 3.4$) and "slow" SiO ($n = 1.65$).

Solution. Since $n = 1.38$ is smaller than the index of the low-index component of the equivalent layer ($n = 1.65$), quarter-wave synthesization is not possible (Fig 3.5). We select a $3\lambda/4$-layer composed of two $3\lambda/8$-layers. Ohmer's equations (Eqs. 3.15 and 3.16) are difficult to use in this case because of the high dispersion near the stopband. We calculate by trial and error: $\phi_A = 15°$ and $\phi_B = 45°$. Figure 6.24 shows the resulting design.

Problem 6.2

Why were the augmented stacks of Figs. 6.18 and 6.19 placed between the main stack and the antireflection coating to the incident medium and not on top?

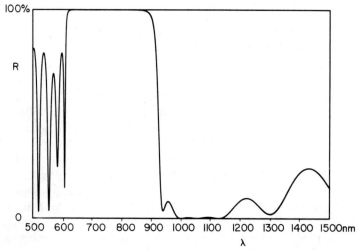

Figure 6.24. Reflectance of the design $1 \mid 1.5(\text{H/6 L1 H/6})^2 (\text{H/2 L2 H/2})^8 \mid 1.52$ with $n_{L1} = 1.65$, $n_{L2} = 1.85$, $n_H = 3.4$, and $\lambda_0 = 740$ nm.

Solution. As seen from the incident-medium side, the stack with the thinner layers and with a rejection band in a region of higher absorption should come first (Sec. 6.3). Also, since the stacks are constructed of the same equivalent layer and the region near the edge is covered up by the rejection band of the existing stack, no additional matching layers or shifted periods are needed.

Problem 6.3

How large is the bandwidth of a stack of alternating layers with n_L, ϕ_L and n_H, ϕ_H in general and specifically for $\phi_L = \phi_H$ when $n_L = 1.45$ and $n_H = 2.3$, or $n_L = 1.85$ and $n_H = 4.2$?

Solution. The characteristic matrix of two equally thick layers with index n_L and n_H is

$$
\mathbf{M} = \begin{bmatrix} \cos\phi_L\cos\phi_H - \dfrac{n_H}{n_L}\sin\phi_L\sin\phi_H & i\dfrac{(\cos\phi_L\sin\phi_H)}{n_H} + i\dfrac{\sin\phi_L\cos\phi_H}{n_L} \\[2ex] i(n_L\sin\phi_L\cos\phi_H + n_H\cos\phi_L\sin\phi_H) & \cos\phi_L\cos\phi_H - \dfrac{n_L}{n_H}\sin\phi_L\sin\phi_H \end{bmatrix}
$$

With Eq. 2.30

$$
\mathbf{M}^p = [S_{p-1}(x)]\mathbf{M} - [S_{p-2}(x)]\mathbf{I}
$$

where $x = M_{11} + M_{22}$.

From the definition of Chebyshev polynomials $[S_{p-1}(x) = \sin p\theta/\sin\theta, 2\cos\theta = x]$ it follows that the stopband is characterized by $x > 2$. This condition yields, for the position of the edges,

$$
x = M_{11} + M_{22} = 2\cos\phi_L\cos\phi_H
$$
$$
- \left(\frac{n_H}{n_L} + \frac{n_L}{n_H}\right)\sin\phi_L\sin\phi_H = \pm 2 \quad (6.7)
$$

For $\phi = \phi_L = \phi_H$ (quarter-wave stack)

$$
\sin\phi = \frac{2\sqrt{n_L n_H}}{n_L + n_H} \quad (6.8)
$$

We obtain for $n_L = 1.45$ and $n_H = 2.3$: $\phi = 76.3°$ and $103.7°$ (bandwidth of 30 percent), and for 1.85 and 4.2: $\phi = 67.1°$ and $112.9°$ (bandwidth of 51 percent).

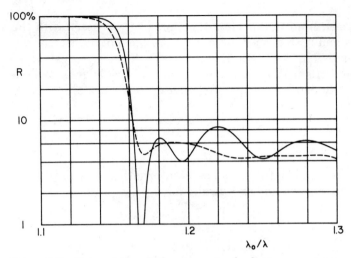

Figure 6.25. Comparison of the two visual short wave passes
1 | 1.12(L/2 H L/2) 1.06(L/2 H L/2) 1.03   (L/2 H L/2) 1.015(L/2 H L/2)
(L/2 H L/2)^5 1.015(L/2 H L/2) 1.03(L/2 H L/2) 1.06(L/2 H L/2) 1.12(L/
2 H L/2) | 1.52 (dashed curve) and 1 | 1.08 (L/2 H L/2) 1.04(L/2 H L/
2) 1.02(L/2 H L/2) 1.01   (L/2 H L/2) (L/2 H L/2)^5 1.01(L/2 H L/2) 1.02 (L/
2 H L/2) 1.04(L/2 H L/2) 1.08(L/2 H L/2) | 1.52   (solid   curve)   with
$n_L = 1.45$ and $n_H = 2.35$.

Problem 6.4

For the visual short wave pass of Fig. 6.12 we used, as a first step, the
relative shift 1.12. How will passband ripple and steepness of slope be
affected if we reduce this first step to 1.08?

Solution.   In Fig. 6.25 we show both designs. The slope between 20 and
90 percent transmittance is decreased from 1.5 to 1.1 percent at a loss
of 2 percent transmittance in the passband. The choice depends obviously on the application.

Minus Filters[†]

Minus filters eliminate one wavelength band from a spectrum. Ideally, the transmittance should be 100 percent from λ_1 to λ_2, the reflectance 100 percent from λ_2 to λ_3, and the transmittance 100 percent again from λ_3 to λ_4. The wavelength region from λ_1 to λ_4 is called the free filter range and the wavelength region from λ_2 to λ_3 the rejection band (stopband, high-reflectance band). For example, the free filter range of a minus green filter would be the visible spectrum, λ_1 = 400 to λ_4 = 700 nm, and the rejection band, the green wavelength region, λ_2 = 500 to λ_3 = 570 nm.

In Chap. 6 we used one side of the stop band of a periodic multilayer to design edge filters. For minus filters we have to use both sides. This makes the design of minus filters considerably more difficult than edge filters. The two edges of a stopband have very different equivalent index characteristics: improving the ripple on one side normally makes the other side worse (so-called "toothpaste tube" effect).

The stopband of periodic multilayers is fairly wide (Prob. 6.3). For a quarter-wave stack with n_L = 1.45 and n_H = 2.3 it is 30 percent. If we use the quarter-wave stack in its third order (as a three-quarter-wave stack) it is 10 percent. Higher orders and ratios of optical thicknesses other than one are possible but not too practical. See Baumeister[2]

---

†From Thelen[1].

for a theoretical study on how narrow a dielectric minus filter might
be possible.

7.1. Quarter-Wave-Stack Minus Filters

From Eqs. 3.1, 3.12, and 3.13 we can derive for the equivalent index
of the structure A/2 B A/2 the relationship

$$\frac{N}{n_A} = \frac{\sqrt{\cos \phi} - (1 - n_B/n_A)/(1 + n_B/n_A)}{\sqrt{\cos \phi} + (1 - n_B/n_A)/(1 + n_B/n_A)} \tag{7.1}$$

and for the equivalent thickness

$$\Gamma = \arccos\left(1 - \frac{(1 + n_B/n_A)^2}{2n_B/n_A}\sin^2 \phi\right) \tag{7.2}$$

$\phi = (\pi/2)(\lambda_0/\lambda)$. Figure 7.1 gives the equivalent index for four ratios
n_B/n_A. From the structure of Eq. 7.1 we see that

$$\frac{N(\lambda_0/\lambda)}{n_A} = \frac{n_A}{N(2 - \lambda_0/\lambda)} \tag{7.3}$$

We have learned before (Eqs. 2.48 and 2.50) that the reflectance and
transmittence of nonabsorbing multilayers are invariant to replacing
all indices by their reciprocal values. We conclude that as long as we
use equivalent structures with an outer common index n_A and as long
as the equivalent structures are centered at the same design wave-

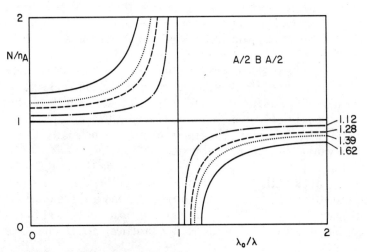

Figure 7.1. Normalized equivalent index N/n_A of the structure A/2 B A/2
for four index ratios n_B/n_A = 1.12, 1.28, 1.39, and 1.62.

length λ_0, all smoothing operations on one side of the stopband $(x = \lambda_0/\lambda)$ are also effective on the other side $(x = 2 - \lambda_0/\lambda)$.

We arrive at the following method of designing minus filters:

1. Select a structure $(A/2\ B\ A/2)^p$ for a multilayer core.

2. Match from n_A (remember that the invariance of the reflectance and transmittance includes the incident-medium and substrate indices) to $N(A/2\ B\ A/2)$ using structures $A/2\ B1\ A/2$, $A/2\ B2\ A/2$, ...

3. Match back from $N(A/2\ B\ A/2)$ to n_A using the same structures $A/2\ B1\ A/2$, $A/2\ B2\ A/2$, ...

4. Match n_A to the incident medium n_0 and substrate n_S.

We call this design method matching with *similar equivalent layers*.

Let us start with single equivalent layer matching. In order to be close enough to the stopband we have to use an equivalent thickness of $3\pi/4$. We generate this thickness by using a $3\pi/8$ layer twice. Let us select $n_A = 1.56$ and $n_B = 2.34$. We use Eq. 7.2 to calculate $\lambda_0/\lambda = 0.72$ with $\Gamma = 3\pi/8$. Equation 7.1 delivers $N(\lambda_0/\lambda = 0.72) = 2.597$, $N = \sqrt{n_A N} = \sqrt{1.56 \times 2.597} = 2.013$. We use Eq. 7.1 again to calculate $n_{B1} = 1.93$ for $N(\lambda_0/\lambda = 0.72) = 2.013$. Figure 7.2 shows the resulting design without matching to the incident medium and substrate. Figure 7.3 gives the design with matching to the massive media and in the third order as a minus green filter. We can use Tables 3.5 to 3.8 to provide more complex matching. In Fig. 7.4 two- and three-layer matching is shown. n_B is now 1.45 and $\lambda_2/\lambda_1 = 6$.

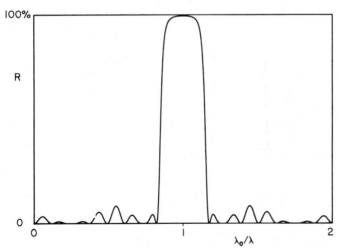

Figure 7.2. Reflectance of the design $1.56 \mid (A/2\ B1\ A/2)^2\ (A/2\ B\ A/2)^6\ (A/2\ B1\ A/2)^2 \mid 1.56$ with $n_A = 1.56$, $nB = 2.34$, and $n_{B1} = 1.93$.

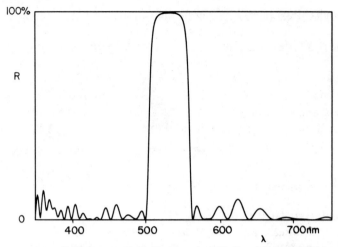

Figure 7.3. Reflectance of the "minus green" filter $1 \mid C1\ 2C2\ (3A/2\ 3B1\ 3A/2)^2\ (3A/2\ 3B\ 3A/2)^6\ (3A/2\ 3B1\ 3A/2)^2 \mid 1.52$ with $n_{C1} = 1.38$, $n_A = 1.56$, $n_{C2} = n_{B1} = 1.93$, and $n_B = 2.34$ ($\lambda_0 = 530$ nm).

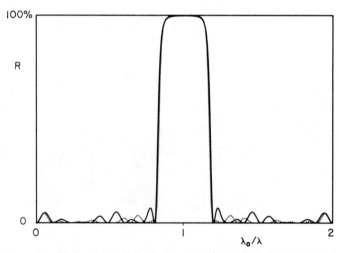

Figure 7.4. Reflectance of the designs $1.45 \mid (A/2\ B1\ A/2)\ (A/2\ B2\ A/2)\ (A/2\ B\ A/2)^6\ (A/2\ B2\ A/2)\ (A/2\ B1\ A/1) \mid 1.45$ with $n_A = 1.45$, $n_{B1} = 1.78$, $n_{B2} = 1.93$, and $n_B = 2.35$ (solid curve) and $1.45 \mid (A/2\ B1\ A/2)\ (A/2\ B2\ A/2)\ (A/2\ B3\ A/2)\ (A/2\ B\ A/2)^6\ (A/2\ B3\ A/2)\ (A/2\ B2\ A/2)\ (A/2\ B1\ A/2) \mid 1.45$ with $n_A = 1.45$, $n_{B1} = 1.67$, $n_{B2} = 1.85$, $n_{B3} = 2.02$, and $n_B = 2.35$ (dotted curve).

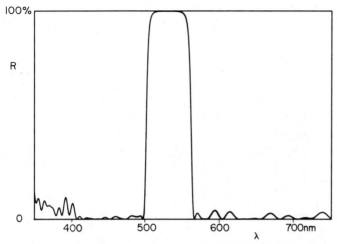

Figure 7.5.   Reflectance of the design 1] C1 2C2 C3 (3A/2 3B1 3A/2) (3A/2 3B2 3A/2) (3A/2 3B3 3A/2) (3A/2 3B 3A/2)^6 (3A/2 3B3 3A/2) (3A/ 2 3B2 3A/2) (3A/2 3B1 3A/2) | 1.52   with $n_A = 1.45$, $n_{B1} = 1.63$, $n_{B2} = 1.85$, $n_{B3} = 2.02$, $n_B = 2.35$, $n_{C1} = 1.38$, $n_{C2} = n_{B3}$,   and $n_{C3} = n_{B1}$ ($\lambda_0 = 530$ nm).

In Fig. 7.5 the design with three-layer matching of Fig. 7.4 is adapted to refractive indices which are more commonly available in the visual.

7.2. Non-Quarter-Wave-Stack Minus Filters

Let us return to Fig. 7.1. When $n_A = n_B$: $N/n_A = 1$ for $0 \leqslant \lambda_0/\lambda < 1$, $N/n_A = \infty$ and 0 for $\lambda_0/\lambda = 1$, and $N/n_A = 1$ again for $1 < \lambda_0/\lambda < 3$. Since we assumed $n_A = n_S = n_0$, we could describe the design of a minus filter as a gradual adaptation from (A/2 Bx A/2) to the core structure (A/2 B A/2), n_{Bx} increasing (or decreasing) in steps from n_A to n_B. The validity of Eq. 7.3 is no longer required. Figure 7.6 gives an example. The equivalent structure is A B A and the adaptation is linear in five steps.

7.3. Equal Ripple Minus Filters

In Fig. 7.7 we show the 31-layer equal ripple prototype filter of Table 3.11 as a minus filter. A shifted "Matched at $\lambda_0/\lambda = 1.2$" version of Table 6.1 was used. The design wavelengths of the antireflection coatings (layers 1, 2, and 62) are 530 nm and the optical thicknesses of the core were multiplied by $\frac{530}{650}$. Should a third-order version be translated into a two-material design each layer of the prototype design has to be synthesized by two or even three equivalent layers! We conclude

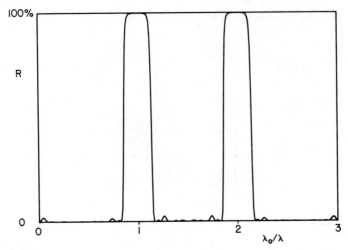

Figure 7.6.   Reflectance of the design 1.45 | 0.667[(A B1 A) (A B2 A) (A B3 A) (A B4 A) (A B5 A) (A B A)^5 (A B5 A) (A B4 A) (A B3 A) (A B2 A) (A B1 A)] | 1.45 with n_A = 1.45, n_{B1} = 1.6, n_{B2} = 1.75, n_{B3} = 1.9, n_{B4} = 2.05, n_{B5} = 2.2, and n_B = 2.35.

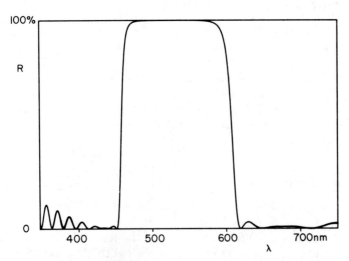

Figure 7.7.   Reflectance of a shifted version of the "Matched at λ_0/λ = 1.2" design of Table 6.1. The optical thicknesses of layers 1, 2, and 62 were multiplied by $\frac{530}{455}$; the thicknesses of all other layers were multiplied by $\frac{530}{650}$.

that unless the stepped refractive indices of the equal ripple prototype designs can be deposited directly, matching with similar equivalent layers (Sec. 7.1) is easier.

7.4. Problems and Solutions

Problem 7.1

For the first step in the design of minus filters by similar equivalent layers the refractive indices of the incident medium, the substrate, and the low-index layer are set equal. Dropping the first and last $\lambda/2$-layer of the design should consequently not matter. Compare the design of Fig. 7.3 with and without outer eight wave layers.

Solution.   Figure 7.8 gives the comparison. It is interesting to note that dropping the eight wave layers restores the symmetry around λ_0.

Problem 7.2

In Sec. 7.2 we had good success with adapting n_B in five steps. Investigate whether linearly adapting the optical thicknesses is equally successful.

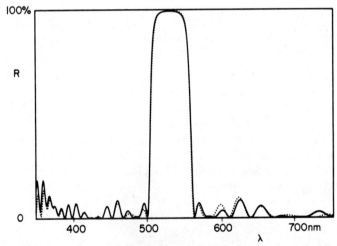

Figure 7.8. Reflectance of the designs $1 \mid C1\ 2C2\ (3B1\ 3A)^2\ (3B\ 3A)^6\ (3B1\ 3A\ 3B1) \mid 1.52$ (solid curve) and $1 \mid C1\ 2C2\ (3A/2\ 3B1\ 3A/2)^2\ (3A/2\ 3B\ 3A/2)^6\ (3A/2\ 3B1\ 3A/2)^2 \mid 1.52$ (dotted curve) with $n_{C1} = 1.38$, $n_A = 1.56$, $n_{C2} = n_{B1} = 1.93$, and $n_B = 2.34$ ($\lambda_0 = 530$ nm).

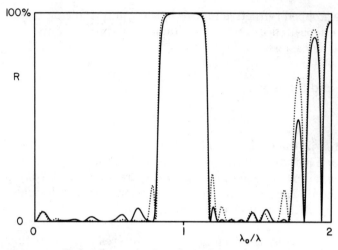

Figure 7.9 Reflectance of the designs 1.45 | (0.898A   0.182B
0.898A) (0.821A 0.326B 0.821A) (0.747A 0.470B 0.747A) (A/2 B A/
2)⁶ (0.747A 0.470B 0.747A) (0.821A 0.326B 0.821A) (0.898A 0.182B
0.898A) | 1.45 (solid curve) and 1.45 | (0.875A 0.25B 0.875A) (0.75A
0.5B 0.75A) (0.625A   0.75B   0.625A) (A/2 B A/2)⁶ (0.625A   0.75B
0.625A) (0.75A 0.5B 0.75A) (0.875A 0.25B 0.875A) | 1.45 (dotted curve)
with $n_A = 1.45$ and $n_B = 2.35$.

Solution. In Fig. 7.9 we compare a linear five-step optical thickness
design with a design derived from the three-equivalent-layer matching
design of Fig. 7.4. Equations 3.15 and 3.16 were used to synthesize the
in-between refractive indices.

Problem 7.3

The method of shifted equivalent layers reduces the ripple near the
edge of an edge filter effectively (Sec. 6.1.3). At the same time it increases the ripple on other edges, though. This is not the case with the
method of matching with similar equivalent indices (Sec. 7.1). Demonstrate.

Solution. In Fig. 7.10 we compare the short-wave-pass design of Fig.
6.10 with the minus filter design of Fig. 7.5. Both designs were shifted
to longer wavelengths. Note that the antireflection coatings were shifted
differently to allow for the wider passband.

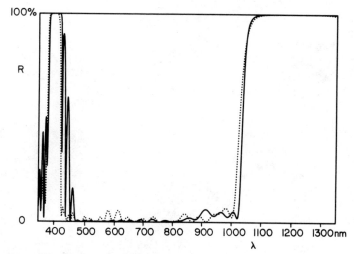

Figure 7.10. Reflectance of the designs $1 \mid 1.05(A/2 \; B \; A/2)^2 \; (A/2 \; B \; A/2)^6 \; 1.05(A/2 \; B \; A/2)^2 \; C1/2 \; C2 \; C3/2 \mid 1.52$ (solid curve) and $1 \mid C1/2 \; C2 \; C3/2 \; (A/2 \; B1 \; A/2) \; (A/2 \; B2 \; A/2) \; (A/2 \; B3 \; A/2) \; (A/2 \; B \; A/2)^6 \; (A/2 \; B3 \; A/2) \; (A/2 \; B2 \; A/2) \; (A/2 \; B1 \; A/2) \mid 1.52$ (dotted curve) with $n_{C1} = 1.38$, $n_A = 1.45$, $n_{C3} = n_{B1} = 1.63$, $n_{B2} = 1.85$, $n_{C2} = n_{B3} = 2.02$, and $n_B = 2.35$ ($\lambda_0 = 1200$ nm).

8

Stopband Suppression and Wide Bandpass Filters

A stopband is, by definition, a wavelength region where the reflection/rejection of a periodic multilayer increases whenever we increase the number of periods p (Sec. 3.1).

8.1. Analog Method of Design[†]

Let $\mathbf{M}^p$ be the characteristic matrix of a periodic multilayer which was generated by repeating a base period p times and let $\mathbf{M}$ be the characteristic matrix of the base period. It then follows (Eq. 2.30)

$$\mathbf{M}^p = [S_{p-1}(x)]\mathbf{M} - [S_{p-2}(x)]\mathbf{I}$$

with $x = M_{11} + M_{22}$. From the definition of the Chebyshev functions S:

$$|x| < 2: \qquad S_{p-1}(x) = \frac{\sin p\theta}{\sin \theta} \qquad 2 \cos \theta = x$$

$$|x| > 2: \qquad S_{p-1}(x) = \frac{\sinh p\theta}{\sinh \theta} \qquad 2 \cosh \theta = x$$

[†]From Epstein.[1]

we can conclude that we have a stopband whenever $|x| > 2$ (same reasoning as in Sec. 3.1.1: trigonometric functions $\rightarrow$ passband and hyperbolic functions $\rightarrow$ stopband).

In Sec. 3.1.2 we established for a basic sequence of layers (the *period*), embedded between equal massive medium ($n_0 = n_S$) (Eq. 3.9):

$$\frac{M_{11} + M_{22}}{2} = \frac{\cos \Phi}{|\vec{T}_e|}$$

Since $|\cos \Phi| \leqslant 1$,

$$|x| = |M_{11} + M_{22}| > 2 \quad \text{only when } |\vec{T}_e| < 1 \quad (8.1)$$

In words: A periodic multilayer can have a stopband only when the transmittance through one period, embedded between equal massive media, is smaller than 100 percent. This is valid for symmetric and nonsymmetric periods and holds true exactly since no approximations were used.

Equation 8.1 is, of course, only a *necessary* and not a *sufficient* condition. For sufficiency, $|\cos \Phi| \approx 1$ (or $\Phi =$ multiple of π) is also required. With approximation Eq. 3.10 ($|\Phi| \approx \Sigma\phi$), this is *approximately* the case when the sum of all phase thicknesses ϕ equals a multiple of π, or

$$\sum_{x=0}^{m} \phi_x = \sum_{x=0}^{m} \left(\frac{2\pi}{\lambda_k}\right) n_x d_x = k\pi \quad \text{with } k = 1, 2, 3, \ldots \quad (8.2)$$

The teachings of Eqs. 8.1 and 8.2 can be combined and expressed in the following way:

A periodic multilayer with symmetric or nonsymmetric periods has generally a stopband when the phase thicknesses of the period add up to a multiple of π and the transmittance through the period, assumed embedded between equal media, is smaller than 100 percent.

Conversely:

The stopband of a periodic multilayer with symmetric or nonsymmetric periods, which normally occurs when the phase thicknesses of the base period add up to a multiple of π, can be suppressed by making the transmittance through the base period, assumed embedded between equal massive media, close to 100 percent.

8.1.1. One stopband suppressed

Let us study a periodic multilayer consisting of periods with two elements (AB). The transmittance through this configuration, assumed embedded between equal massive media, is always less than 100 percent except for the case when the optical thickness of one of the elements is a multiple of a half wave. Optically, a half wave acts as if it were not there (Sec. 3.5). Consequently, one can choose the index of the massive media equal to the index of the other element which results in 100 percent transmittance. After some manipulation we can formulate this result in the following way: A periodic multilayer with two alternating layers A and B has a stopband at

$$\lambda_k = \frac{2(n_A d_A + n_B d_B)}{k} \quad \text{with } k = 1, 2, 3, \ldots \quad (8.3)$$

except when

$$k = p(a + b) \quad \text{with } p = 1, 2, 3, \ldots$$

where a and b are the smallest possible integers defined by the ratio of optical thicknesses $n_A d_A / n_B d_B = a/b$.

Examples are the quarter-wave stack ($n_A d_A = n_B d_B$, every second stopband suppressed, Figs. 3.3 and 3.4) and the quarter/half stack ($n_A d_A = 2n_B d_B$ or $2n_A d_A = n_B d_B$, every third stopband suppressed, Fig. 3.6). As we discussed before (Sec. 2.11), from a periodicity point of view designs with periods (A B) are equal to (A/2 B A/2) and designs with periods (A B B) are equal to (A/2 BB A/2) or (B A B).

8.1.2. Two stopbands suppressed[†]

In order to suppress two stopbands we consider a period made of two back-to-back two-layer antireflection coatings (Epstein[1]):

$$\text{(aA bB cC bB aA)} \quad (8.4)$$

aA bB acts as a two-layer antireflection coating from an embedding medium E (called the *dummy* medium since its actual presence is not required in the final multilayer) to the medium C. For the purpose of our calculation we can assume medium C to be a massive medium since at the wavelength positions of zero reflectance it no longer acts as an interference film (no zigzag reflections).

E | aA bB | C is an antireflection coating with two minima ($\lambda_{k1}/\lambda_{k2} < 3$) when (Collin's formulas, Sec. 3.2) $a = b = 1$ (or $n_A d_A = n_B d_B$) and

$$n_A n_B = n_E n_C \quad (8.5)$$

The location of the points of zero reflectance are determined by Eq. 4.8. For the simultaneous suppression of two stopbands Eq. 4.8 has to

---

[†]From Epstein[1] and Thelen.[2]

be satisfied at both locations λ_{k1} and λ_{k2}. This is not difficult since the tangent functions of Eq. 4.8 appear to the second power, and for each solution ϕ there is a solution $\pi = \phi$ [$\tan \phi = -\tan(\pi - \phi)$]. We can consequently write

$$\left(\frac{2\pi}{\lambda_{k1}}\right) n_A d_A = \phi \quad \text{or} \quad = \pi - \phi$$

and

$$\left(\frac{2\pi}{\lambda_{k2}}\right) n_A d_A = \pi - \phi \quad \text{or} \quad = \phi$$ (8.6)

or

$$\phi = \frac{\pi}{1 + \lambda_{k1}/\lambda_{k2}} \quad \text{or} \quad \phi = \frac{\pi}{1 + \lambda_{k2}/\lambda_{k1}}$$

Since the lower order must be connected with the smaller phase angle (longer wavelength) the second alternative is valid

$$\phi_{Ak1} = \frac{\pi}{1 + \lambda_{k2}/\lambda_{k1}} = \frac{\pi}{1 + k_2/k_1}$$ (8.7)

where k_1 and k_2 stand for the orders of the suppressed bands.

If we insert this value into Eq. 4.8, set $n_1 = n_A$, $n_2 = n_B$, $n_S = n_C$, and eliminate $n_M = n_0$ we obtain, with Eq. 8.5,

$$\tan^2 \frac{\pi}{1 + \lambda_{k1}/\lambda_{k2}} = \frac{n_A n_B - n_C^2}{n_B^2 - n_A n_C^2/n_B}$$ (8.8)

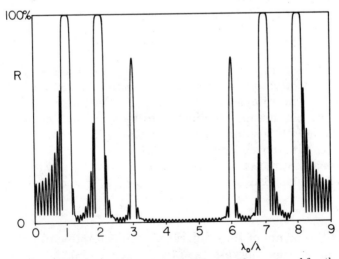

Figure 8.1. Reflectance of a periodic multilayer with suppressed fourth and fifth orders: $1.0 \mid 0.222[(\text{L M 5H M L})^{10} \text{ L}] \mid 1.52$ with $n_L = 1.38$, $n_M = 1.926$, and $n_H = 2.3$.

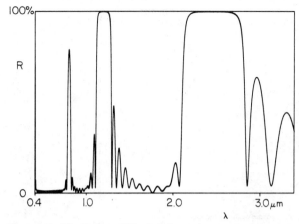

Figure 8.2.   The design of Fig. 8.1 shown in wavelengths with $\lambda_0 = 0.533$ μm. Reflectance of $1 \mid (\text{L M 5H M L})^{10} \text{ L} \mid 1.52$ with $n_L = 1.38$, $n_M = 1.926$, $n_H = 2.3$, and $\lambda_0 = 0.533$ μm.

With Eq. 8.8 we can calculate for every two given refractive indices the third. Let us remember, though, that we used an approximation (Eq. 3.10) in the derivation of Eq. 8.8. As a consequence, the accuracy of refractive indices calculated with Eq. 8.8 is limited.

The optical thicknesses are determined with Eqs. 8.2, 8.5, and 2.5. For $k = k_1$ (location of the first suppressed stopband), Eq. 8.2 delivers

$$2\phi_{Ak1} + 2\phi_{Bk1} + \phi_{Ck1} = k_1 \pi \qquad \text{or} \qquad \phi_{Ck1} = k_1 \pi - 4\phi_{Ak1} \qquad (8.9)$$

In his pioneering paper on order suppression Epstein[1] gave a design with the fourth and fifth order suppressed: $k_1 = 4$, $\phi_{Ak1} = 80°$ (Eq. 8.7), and $\phi_{Ck1} = 400°$ (Eq. 8.9). With $n_A = 1.38$ and $n_C = 2.3$ we determine from Eq. 8.8 by iteration $n_B = 1.926$. Figures 8.1 and 8.2 give the design with ten periods in wavenumber and wavelength.

Similarly, Thelen[2] designed a filter with the second and third orders suppressed: $k_1 = 2$, $\phi_{Ak1} = \phi_{Ck1} = 72°$. For $n_A = 1.38$ and $n_C = 2.3$, n_B turns out to be $n_B = 1.90$. Figure 8.3 gives the design with ten periods in wavenumbers. In Fig. 8.4 the application as *red and blue solar cell reflector* is shown and in Fig. 8.5 as *double-stack heat reflector* (Thelen[2]).

8.1.3. Three adjacent orders suppressed†

There is a unique solution to Eq. 8.7 when $k_2/k_1 = {}^3/_1$. Then $\phi_{Ak1} = \phi_{Bk1}$ $= 90°$, $\phi_{Ck1} = 180°$, and $n_B = \sqrt{n_A n_C}$ (Eq. 8.8). In spite of the fact that we initially only eliminated the second and fourth order, the third is also absent since at this position all layers are multiples of half waves

---

†From Thelen.[2]

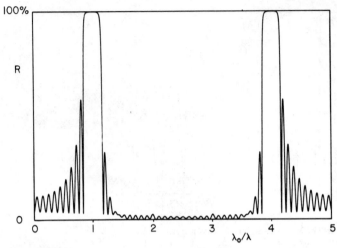

Figure 8.3. Reflectance of a periodic multilayer with suppressed second and third orders: $1.0 \mid 0.4[(LMHML)^{10}L] \mid 1.50$ with $n_L = 1.38$, $n_M = 1.90$, and $n_H = 2.3$.

and consequently "absentee"-layers (Sec. 3.5). Figure 8.6 gives an example with ten periods and Fig. 8.7 a *triple-stack heat reflector* (Thelen[2]).

8.1.4. Order suppression with two coating materials

Although Eq. 8.8 is quite tolerant to index variations and an exact match of the indices of the three coating materials is consequently not required, designs with two coating materials would be preferable.

The in-between refractive index could easily be synthesized (Eqs. 3.15 and 3.16). But this alone does not solve the problem since the synthesization is only valid for a narrow wavelength region and we require for two suppressed orders a good match at two wavelength positions. One way to solve the problem is to synthesize the in-between index as a first design step and use refining techniques (Chap. 11) to bring about "zero" reflectance at the two required wavelength positions. Problem 8.2 gives an example. The structure of the base period is

$$(a_1A \; b_1B \; a_2A \; b_2B \; a_2A \; b_1B \; a_1A) \qquad (8.10)$$

8.2. Analytical Method of Design[†]

In the derivation of Eq. 8.2 the assumption was made that the multiple (zigzag) reflections inside the base period can be neglected. This, of

---

[†]From Thelen.[3]

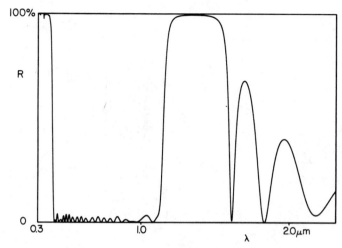

Figure 8.4. Reflectance of a "red and blue solar cell reflector":
1.0 | 1.05(LMHML) (LMHML)^6 1.15(LMHML) 1.55L 0.628 (H/2 L H/
2)^6 0.565(H/2 L H/2)^2 | 1.50 with $n_L = 1.38$, $n_M = 1.90$, and $n_H = 2.3$
($\lambda_0 = 525$ nm).

course, is only approximately true. We now develop formulas for the
exact suppression of stopbands.

In Sec. 3.1 we defined the equivalent index of a symmetric period as
(Eq. 3.1)

$$N = + \sqrt{\frac{M_{21}}{M_{12}}}$$

If we restrict ourselves to an uneven number m of equally thick layers
with equal phase thickness ϕ we have, with Eq. 3.21,

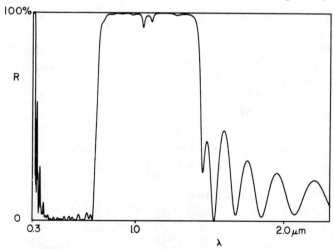

Figure 8.5. Reflectance of a "double-stack heat reflector": 1.0 |
1.1(L/2 H L/2) (L/2 H L/2)^5 1.125 (L/2 H L/2) 0.57(LMHML)^8 L/2 | 1.50
with $n_L = 1.38$, $n_M = 1.90$, and $n_H = 2.3$ ($\lambda_0 = 0.860$ μm).

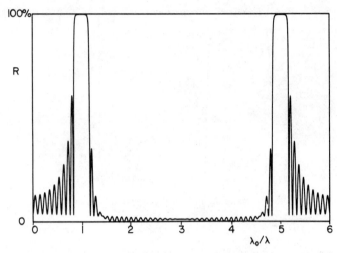

Figure 8.6. Reflectance of the periodic multilayer with suppressed second, third, and fourth orders: 1.0 | 0.3333[(LM 2H M L)^{10} L] | 1.50 with $n_L = 1.38$, $n_M = 1.781$, and $n_H = 2.3$.

$$N = \sqrt{\frac{a_1 \cos^{m-1}\phi + a_3\cos^{m-3}\phi + \cdots}{b_1 \cos^{m-1}\phi + b_3\cos^{m-3}\phi + \cdots}} \qquad (8.11)$$

Equation 8.11 is a ratio of two entire functions in $\cos\phi$ and can consequently always be brought into the following form:

$$N = C\sqrt{\frac{(\cos\phi - \alpha_1)(\cos\phi - \alpha_2)\cdots}{(\cos\phi - \beta_1)(\cos\phi - \beta_2)\cdots}} \qquad (8.12)$$

When $\cos\phi = \alpha_x$ the radicand changes sign. This means we are at the edge of a stopband. The same is true when $\cos\phi = \beta_y$.

Now, if we equate α_x and β_y, we can cancel $(\cos\phi - \alpha_x)$ against $(\cos\phi - \beta_y)$ in Eq. 8.12 and a periodic multilayer with this base period will have one stopband less than before.

8.2.1. Periodic multilayers (ABCBA)^p with two suppressed stopbands

For the structure (ABCBA) we calculate, by straightforward matrix multiplication and after some trigonometric manipulations,

$$N^2 = n_A^2 \frac{\cos^2\phi + [1/(1 + n_C/n_B)]\cos\phi - \{1/[(1 + n_B/n_A)(1 + n_B/n_C)]\}}{\cos^2\phi + [1/(1 + n_B/n_C)]\cos\phi - \{1/[(1 + n_A/n_B)(1 + n_C/n_B)]\}}$$

$$\frac{\cos^2\phi - [1/(1 + n_C/n_B)]\cos\phi - \{1/[(1 + n_B/n_A)(1 + n_B/n_C)]\}}{\cos^2\phi - [1/(1 + n_B/n_C)]\cos\phi - \{1/[(1 + n_A/n_B)(1 + n_C/n_B)]\}}$$

$$(8.13)$$

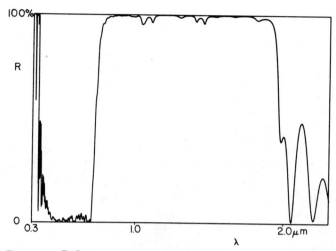

Figure 8.7. Reflectance of the "triple-stack heat reflector": 1.0 | 1.1
(L/2 H L/2) (L/2 H L/2)^5 1.25(L/2 H L/2) 0.57(L M1 H M1 L)^8 0.642(L
M2 2H M2 L)^8 L/2 | 1.5 with n_L = 1.38, n_{M1} = 1.90, n_{M2} = 1.781,
and n_H = 2.3 (λ_0 = 0.860 μm).

Setting $x = n_B/n_C$, $y = n_A/n_B$, bringing the equation to the form of Eq.
8.12, and equating the α's and β's of the first and second fraction leads
to

$$\frac{1}{2}\frac{x}{1+x} \pm \left[\frac{1}{4}\frac{x^2}{(1+x)^2} + \frac{y}{(1+x)(1+y)}\right]^{1/2}$$

$$= \frac{1}{2}\frac{1}{1+x} \pm \left[\frac{1}{4}\frac{1}{(1+x)^2} + \frac{x}{(1+x)(1+y)}\right]^{1/2}.$$

By isolating one square root at one side, squaring the equation, isolating the remaining square root at one side, and squaring again, we
obtain

$$y^2 - \frac{(1+x)(1+x^2)}{2x}y + x = 0 \tag{8.14}$$

with the solutions

$$y_1 = \frac{(1+x)(1+x^2)}{4x} + \left[\frac{(1+x)^2(1+x^2)^2}{16x^2} - x\right]^{1/2} \tag{8.15}$$

and $\quad y_2 = \frac{(1+x)(1+x^2)}{4x} - \left[\frac{(1+x)^2(1+x^2)^2}{16x^2} - x\right]^{1/2} \tag{8.16}$

Periodic multilayers with base structures according to solution y_1 have the second and third order suppressed, and those according to solutions y_2 have the first and fourth order suppressed. In Fig. 8.8 we compare the two solutions with a quarter/quarter (Sec. 3.1.1.) and a half/quarter structure (Sec. 3.1.5).

Tables 8.1 and 8.2 give numeric values for the solutions of Eqs. 8.15 and 8.16. The tables are arranged so that the ratio of the highest to the lowest refractive index is the independent variable and the ratio of the in-between index to the lowest index is the dependent variable. For the calculation of Table 8.1 we transformed Eq. 8.14 to a third-order equation in $n_C/n_A = 1/xy$ and $n_B/n_A = 1/y$ (see Prob. 8.2). Table 8.2 was easier to calculate since now n_B is the highest and n_C the lowest index ($n_B/n_C = x$ and $n_A/n_C = xy$).

Equation 8.8 should be identical to Eq. 8.15 when $\lambda_{k1}/\lambda_{k2} = {}^3/_2$ and to Eq. 8.16 when $\lambda_{k1}/\lambda_{k2} = {}^5/_4$. They obviously are not. But for $x = n_B/n_C = 1 + \varepsilon$ and $\lambda_{k1}/\lambda_{k2} = {}^2/_3$, Eqs. 8.8 and 8.15 can both be approximated by $y = n_A/n_B = 1 + 1.618\varepsilon$. For $\lambda_{k1}/\lambda_{k2} = {}^5/_4$ and $x = n_B/n_C = 1 + \varepsilon$ Eq. 8.8 and 8.16 can be approximated by $y = n_A/n_B = 1 + 0.618\varepsilon$. We conclude that the analog design method of Sec. 8.1.2 delivers a first-order approximation to the exact solutions.

As an application of a periodic multilayer with suppressed first and fourth orders we show it in Fig. 8.9 in combination with a quarter-

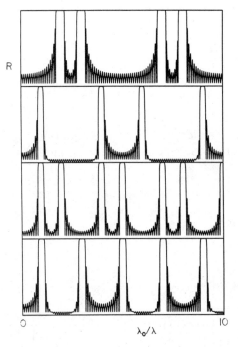

R

0   λ_0/λ I0

Figure 8.8. Stopband suppression in periodic multilayers (from bottom to top): $1 \mid (A/2\ C\ A/2)^{10} \mid 1.52$, $1 \mid 0.66667(ACA)^{10} \mid 1.52$, $1 \mid 0.4(ABCBA)^{10} \mid 1.52$, and $1 \mid 0.4(DCACD)^{10} \mid 1.52$ with $n_A = 1.45$, $n_B = 1.955$, $n_C = 2.35$, and $n_D = 1.736$.

TABLE 8.1   Relative Refractive Index n_B/n_A as a Function of n_C/n_A of a Structure (ABCBA) which Generates Suppressed Second and Third Orders If Used as a Base Period of a Periodic Multilayer

| | n_B/n_A | | | | | | | | |
n_C/n_A	0.00	0.01	0.02	0.03	0.04	0.05	0.06	0.07	0.08	0.09
1.00	1.0000	1.0062	1.0123	1.0184	1.0245	1.0306	1.0367	1.0427	1.0487	1.0547
1.10	1.0607	1.0666	1.0726	1.0785	1.0844	1.0902	1.0961	1.1019	1.1077	1.1135
1.20	1.1193	1.1250	1.1308	1.1365	1.1422	1.1479	1.1536	1.1592	1.1648	1.1705
1.30	1.1761	1.1816	1.1872	1.1928	1.1983	1.2038	1.2093	1.2148	1.2203	1.2257
1.40	1.2312	1.2366	1.2420	1.2474	1.2528	1.2582	1.2636	1.2689	1.2742	1.2795
1.50	1.2849	1.2901	1.2954	1.3007	1.3059	1.3112	1.3164	1.3216	1.3268	1.3320
1.60	1.3372	1.3423	1.3475	1.3526	1.3578	1.3629	1.3680	1.3731	1.3782	1.3832
1.70	1.3883	1.3933	1.3984	1.4034	1.4084	1.4134	1.4184	1.4234	1.4283	1.4333
1.80	1.4383	1.4432	1.4481	1.4530	1.4580	1.4629	1.4677	1.4726	1.4775	1.4824
1.90	1.4872	1.4920	1.4969	1.5017	1.5065	1.5113	1.5161	1.5209	1.5257	1.5304
2.00	1.5352	1.5399	1.5447	1.5494	1.5541	1.5588	1.5635	1.5682	1.5729	1.5776
2.10	1.5823	1.5869	1.5916	1.5962	1.6009	1.6055	1.6101	1.6147	1.6193	1.6239
2.20	1.6285	1.6331	1.6377	1.6423	1.6468	1.6514	1.6559	1.6604	1.6650	1.6695
2.30	1.6740	1.6785	1.6830	1.6875	1.6920	1.6965	1.7009	1.7054	1.7099	1.7143
2.40	1.7187	1.7232	1.7276	1.7320	1.7364	1.7409	1.7453	1.7496	1.7540	1.7584
2.50	1.7628	1.7672	1.7715	1.7759	1.7802	1.7846	1.7889	1.7932	1.7976	1.8019
2.60	1.8062	1.8105	1.8148	1.8191	1.8234	1.8277	1.8319	1.8362	1.8405	1.8447
2.70	1.8490	1.8532	1.8574	1.8617	1.8659	1.8701	1.8743	1.8786	1.8828	1.8870
2.80	1.8912	1.8953	1.8995	1.9037	1.9079	1.9120	1.9162	1.9204	1.9245	1.9286
2.90	1.9328	1.9369	1.9410	1.9452	1.9493	1.9534	1.9575	1.9616	1.9657	1.9698

wave stack. The stopband of the quarter-wave stack just fits into the passband between the second- and third-order stopbands of the stack with suppressed orders.

8.2.2.   Periodic multilayers (ABCDDCBA)^p with five suppressed stopbands

Let us consider the structure

$$(ABCDDCBA) \qquad (8.17)$$

Again, we calculate the equivalent index N and equate the α's and β's. As a first condition we find

$$n_A n_D = n_B n_C \qquad (8.18)$$

and as a second condition

$$y - 2y \frac{x^2 - x + 1}{x} + 1 = 0 \qquad (8.19)$$

TABLE 8.2. Relative Refractive Index n_A/n_C as a Function of n_B/n_C of a Structure (ABCBA) which Generates Suppressed First and Fourth Orders If Used as a Base Period of a Periodic Multilayer

	n_A/n_C									
n_B/n_C	0.00	0.01	0.02	0.03	0.04	0.05	0.06	0.07	0.08	0.09
1.00	1.0000	1.0038	1.0076	1.0114	1.0151	1.0188	1.0225	1.0262	1.0298	1.0334
1.10	1.0370	1.0406	1.0442	1.0477	1.0512	1.0547	1.0582	1.0617	1.0651	1.0685
1.20	1.0719	1.0753	1.0786	1.0819	1.0852	1.0885	1.0918	1.0950	1.0983	1.1015
1.30	1.1047	1.1078	1.1110	1.1141	1.1172	1.1203	1.1234	1.1265	1.1295	1.1325
1.40	1.1356	1.1385	1.1415	1.1445	1.1474	1.1503	1.1532	1.1561	1.1590	1.1618
1.50	1.1647	1.1675	1.1703	1.1731	1.1758	1.1786	1.1813	1.1841	1.1868	1.1895
1.60	1.1921	1.1948	1.1974	1.2001	1.2027	1.2053	1.2079	1.2104	1.2130	1.2155
1.70	1.2181	1.2206	1.2231	1.2256	1.2280	1.2305	1.2329	1.2354	1.2378	1.2402
1.80	1.2426	1.2450	1.2473	1.2497	1.2520	1.2544	1.2567	1.2590	1.2613	1.2635
1.90	1.2658	1.2681	1.2703	1.2725	1.2747	1.2769	1.2791	1.2813	1.2835	1.2856
2.00	1.2878	1.2899	1.2920	1.2942	1.2963	1.2983	1.3004	1.3025	1.3045	1.3066
2.10	1.3086	1.3107	1.3127	1.3147	1.3167	1.3186	1.3206	1.3226	1.3245	1.3265
2.20	1.3284	1.3303	1.3322	1.3342	1.3360	1.3379	1.3398	1.3417	1.3435	1.3454
2.30	1.3472	1.3490	1.3508	1.3527	1.3545	1.3563	1.3580	1.3598	1.3616	1.3633
2.40	1.3651	1.3668	1.3685	1.3703	1.3720	1.3737	1.3754	1.3771	1.3787	1.3804
2.50	1.3821	1.3837	1.3854	1.3870	1.3887	1.3903	1.3919	1.3935	1.3951	1.3967
2.60	1.3983	1.3999	1.4014	1.4030	1.4045	1.4061	1.4076	1.4092	1.4107	1.4122
2.70	1.4137	1.4152	1.4167	1.4182	1.4197	1.4212	1.4227	1.4241	1.4256	1.4270
2.80	1.4285	1.4299	1.4313	1.4328	1.4342	1.4356	1.4370	1.4384	1.4398	1.4412
2.90	1.4425	1.4439	1.4453	1.4466	1.4480	1.4493	1.4507	1.4520	1.4534	1.4547

with the solutions

$$y_1 = \frac{x^2 - x + 1}{x} + \frac{x - 1}{x} \sqrt{1 + x^2} \tag{8.20}$$

and

$$y_2 = \frac{x^2 - x + 1}{x} - \frac{x - 1}{x} \sqrt{1 + x^2} \tag{8.21}$$

x and y are defined as $x = n_A/n_B$ and $y = n_B/n_C$. Figure 8.10 gives a plot of one example each of the types of solutions. y_1 has the second through sixth order suppressed and y_2 the first, second, fourth, sixth, and seventh. Tables 8.3 and 8.4 give solutions for designs according to y_1 and Tables 8.5 and 8.6 for designs according to y_2. The ratios of the highest to the lowest refractive indices are taken as independent variables and the ratio of the in-between indices to the lowest index as dependent variables.

In Fig. 8.11 we show an infrared reflector with high reflectance in the 2.5- to 3-μm region and high transmittance all the way to below the visual region based on a solution y_1.

8.3.   Problems and Solutions

Problem 8.1

Verify the conclusions of Sec. 8.1 by analytically determining x for a periodic multilayer with two layers in the base period.

Solution. For two layers with indices n_L, n_H and phase thicknesses ϕ_L, ϕ_H we can deduce from Eqs. 2.16 and 2.21:

$$
\mathbf{M}_{LH} = \begin{bmatrix} \cos \phi_L & \dfrac{i \sin \phi_L}{n_L} \\ i n_L \sin \phi_L & \cos \phi_L \end{bmatrix} \begin{bmatrix} \cos \phi_H & \dfrac{i \sin \phi_H}{n_H} \\ i n_H \sin \phi_H & \cos \phi_H \end{bmatrix}
$$

$$
= \begin{bmatrix} \cos \phi_L \cos \phi_H - \dfrac{n_H \sin \phi_L \sin \phi_H}{n_L} & i\left(\dfrac{\cos \phi_L \sin \phi_H}{n_H} + \dfrac{\sin \phi_L \cos \phi_H}{n_L} \right) \\ i(n_L \sin \phi_L \cos \phi_H + n_H \cos \phi_L \sin \phi_H) & -\dfrac{n_L \sin \phi_L \sin \phi_H}{n_H} + \cos \phi_L \cos \phi_H \end{bmatrix}
$$

$$
x = M_{11} + M_{22} = 2 \cos \phi_L \cos \phi_H - \left(\frac{n_H}{n_L} + \frac{n_L}{n_H} \right) \sin \phi_L \sin \phi_H
$$

$$
= \frac{1}{2} \left(2 - \frac{n_L}{n_H} - \frac{n_H}{n_L} \right) \cos (\phi_L + \phi_H) + \frac{1}{2} \left(2 + \frac{n_L}{n_H} + \frac{n_H}{n_L} \right) \cos (\phi_L - \phi_H)
$$

$$
= \cos (\phi_L - \phi_H) + \cos (\phi_L + \phi_H) + \frac{1}{2} \left(\frac{n_L}{n_H} + \frac{n_H}{n_L} \right) [\cos (\phi_L - \phi_H) - \cos (\phi_L + \phi_H)]
$$

$$
\tag{8.22}
$$

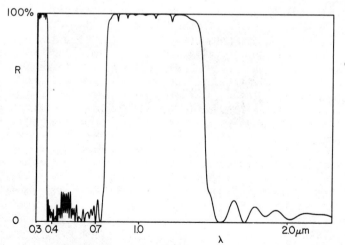

Figure 8.9. Reflectance of the design $1 \mid 3A \, (BACAB)^5 \, AB \, (AC)^5 \, AB \mid 1.52$ with $n_A = 1.46$, $n_B = 1.99$, and $n_C = 2.393$ ($\lambda_0 = 1 \, \mu m$).

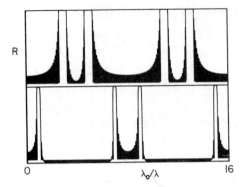

Figure 8.10. Suppression of five adjacent stopbands: $1 \mid 0.25(ABCDDCBA)^{10} \mid 1.52$ with $n_A = 1.45$, $n_B = 1.67$, $n_C = 2.04$, and $n_D = 2.35$ (bottom); also $1 \mid 0.25(ABCDDCBA)^{10} \mid 1.52$ with $n_A = 1.668$, $n_B = 2.35$, $n_C = 1.45$, and $n_D = 2.043$ (top).

For $\phi_L = \phi_H$ (quarter-wave stack) $|x| = n_L/n_H + n_L/n_H > 2$ (stopbands) when $\phi_L + \phi_H = \pi, 3\pi, 5\pi, \ldots$ and $|x| = 2$ (no stopband) when $\phi_L + \phi_H = 2\pi, 4\pi, \ldots$, assuming, of course, that $n_L/n_H \neq 1$. This verifies our previous observation that in the spectrum of a quarter-wave stack every second possible stopband is suppressed.

TABLE 8.3. Relative Refractive Indices n_B/n_A, n_C/n_A as a Function of n_D/n_A of a Structure (ABCDDCBA) which Generates Suppressed Second, Third, Fourth, Fifth, and Sixth Orders If Used as a Base Period of a Periodic Multilayer†

n_D/n_A	n_B/n_A 0.00	0.01	0.02	0.03	0.04	0.05	0.06	0.07	0.08	0.09
1.00	1.0000	1.0029	1.0058	1.0087	1.0116	1.0144	1.0172	1.0200	1.0228	1.0256
	1.0000	1.0071	1.0141	1.0211	1.0281	1.0351	1.0421	1.0490	1.1559	1.0628
1.10	1.0283	1.0310	1.0337	1.0364	1.0391	1.0481	1.0444	1.0471	1.0497	1.0503
	1.0697	1.0766	1.0834	1.0903	1.0971	1.1039	1.1106	1.1174	1.1242	1.1309
1.20	1.0549	1.0574	1.0600	1.0625	1.0650	1.0675	1.0700	1.0725	1.0750	1.0774
	1.1376	1.1443	1.1510	1.1576	1.1643	1.1709	1.1775	1.1841	1.1907	1.1973
1.30	1.0799	1.0823	1.0847	1.0871	1.0895	1.0919	1.0943	1.0966	1.0989	1.1013
	1.2038	1.2104	1.2169	1.2234	1.2299	1.2364	1.2429	1.2493	1.2557	1.2622
1.40	1.1036	1.1059	1.1082	1.1105	1.1127	1.1150	1.1172	1.1195	1.1217	1.1239
	1.2686	1.2750	1.2814	1.2877	1.2941	1.3004	1.3068	1.3131	1.3194	1.3257
1.50	1.1261	1.1283	1.1305	1.1327	1.1348	1.1370	1.1391	1.1413	1.1434	1.1455
	1.3320	1.3383	1.3445	1.3508	1.3570	1.3632	1.3694	1.3756	1.3818	1.3880
1.60	1.1476	1.1497	1.1518	1.1539	1.1560	1.1580	1.1601	1.1621	1.1642	1.1662
	1.3942	1.4003	1.4065	1.4126	1.4187	1.4248	1.4309	1.4370	1.4431	1.4491
1.70	1.1682	1.1702	1.1722	1.1742	1.1762	1.1782	1.1802	1.1812	1.1841	1.1860
	1.4552	1.4612	1.4673	1.4733	1.4793	1.4853	1.4913	1.4973	1.5033	1.5092
1.80	1.1800	1.1899	1.1918	1.1937	1.1957	1.1976	1.1995	1.2013	1.2032	1.2051
	1.5152	1.5211	1.5271	1.5330	1.5389	1.5448	1.5507	1.5566	1.5625	1.5683
1.90	1.2070	1.2088	1.2107	1.2152	1.2144	1.2162	1.2180	1.2198	1.2217	1.2235
	1.5742	1.5800	1.5859	1.5917	1.5975	1.6034	1.6092	1.6150	1.6207	1.6265

†$1.00 \leq n_D/n_A \leq 1.99$

TABLE 8.4. Relative Refractive Indices n_B/n_A, n_C/n_A as a Function of n_D/n_A of a Structure (ABCDDCBA) which Generates Suppressed Second, Third, Fourth, Fifth, and Sixth Orders If Used as a Base Period of a Periodic Multilayer†

	n_B/n_A n_C/n_A								
n_D/n_A	0.00	0.01	0.02	0.03	0.04	0.05	0.06	0.07	0.08	0.09
2.00	1.2253	1.2271	1.2289	1.2306	1.2324	1.2342	1.2359	1.2377	1.2395	1.2412
	1.6323	1.6381	1.6438	1.6495	1.6553	1.6610	1.6667	1.6742	1.6781	1.6838
2.10	1.2429	1.2447	1.2464	1.2481	1.2499	1.2516	1.2533	1.2550	1.2567	1.2584
	1.6895	1.6952	1.7009	1.7065	1.7122	1.7179	1.7235	1.7291	1.7347	1.7404
2.20	1.2600	1.2617	1.2634	1.2651	1.2667	1.2684	1.2700	1.2717	1.2733	1.2750
	1.7460	1.7516	1.7572	1.7628	1.7683	1.7739	1.7795	1.7850	1.7906	1.7961
2.30	1.2766	1.2782	1.2799	1.2815	1.2831	1.2847	1.2863	1.2879	1.2895	1.2911
	1.8017	1.8072	1.8127	1.8182	1.8237	1.8292	1.8347	1.8402	1.8457	1.8512
2.40	1.2927	1.2942	1.2958	1.2974	1.2990	1.3005	1.3021	1.3036	1.3052	1.3067
	1.8566	1.8621	1.8675	1.8730	1.8784	1.8839	1.8893	1.8947	1.9001	1.9055
2.50	1.3083	1.3098	1.3113	1.3129	1.3144	1.3159	1.3174	1.3189	1.3204	1.3219
	1.9109	1.9163	1.9217	1.9271	1.9325	1.9378	1.9432	1.9485	1.9539	1.9592
2.60	1.3234	1.3249	1.3264	1.3279	1.3294	1.3309	1.3324	1.3338	1.3353	1.3368
	1.9646	1.9699	1.9752	1.9805	1.9859	1.9912	1.9965	2.0018	2.0071	2.0123
2.70	1.3382	1.3397	1.3411	1.3426	1.3440	1.3455	1.3469	1.3483	1.3498	1.3512
	2.0176	2.0229	2.0282	2.0334	2.0387	2.0439	2.0492	2.0544	2.0596	2.0649
2.80	1.3526	1.3540	1.3554	1.3569	1.3583	1.3597	1.3611	1.3625	1.3639	1.3653
	2.0701	2.0753	2.0805	2.0857	2.0909	2.0961	2.1013	2.1065	2.1116	2.1168
2.90	1.3666	1.3680	1.3694	1.3708	1.3722	1.3735	1.3749	1.3763	1.3776	1.3790
	2.1220	2.1271	2.1323	2.1374	2.1426	2.1477	2.1529	2.1580	2.1631	2.1682

†$2.00 \le n_D/n_A \le 2.99$

For non-quarter-wave stacks, we have a stopband whenever $\cos(\phi_L - \phi_H) - \cos(\phi_L + \phi_H) = \pm 2$ and a suppressed stopband whenever $\cos(\phi_L - \phi_H) - \cos(\phi_L + \phi_H) = 0$. For a half/quarter stack there are consequently stopbands at the first- and second-, but not at the third-order position, and for a 3:1 stack at the first-, second-, and third-, but not at the fourth-order position, etc.

Problem 8.2

Transform Eq. 8.14 into an equation in $u = n_B/n_A$ and $v = n_C/n_A$ instead of $x = n_B/n_C$ and $y = n_A/n_B$.

Solution. With $x = u/v$ and $y = 1/u$ we obtain, from Eq. 8.14,

$$(2v - 1)u^3 - vu^2 - v^2u + v^2(2 - v) = 0 \tag{8.23}$$

or

$$v^3 + (u - 2)v^2 + u^2(1 - 2u)v + u^3 = 0 \tag{8.24}$$

TABLE 8.5.   Relative Refractive Indices n_A/n_C, n_D/n_C as a Function of n_B/n_C of a Structure (ABCDDCBA) which Generates Suppressed First, Second, Fourth, Fifth, Sixth, and Seventh Orders If Used as a Base Period of a Periodic Multilayer†

n_B/n_C	n_A/n_C n_D/n_C									
	0.00	0.01	0.02	0.03	0.04	0.05	0.06	0.07	0.08	0.09
1.00	1.0000	1.0029	1.0058	1.0087	1.0116	1.0144	1.0172	1.0200	1.0228	1.0256
	1.0000	1.0071	1.0141	1.0211	1.0281	1.0351	1.0421	1.0490	1.0559	1.0628
1.10	1.0283	1.0310	1.0337	1.0364	1.0391	1.0471	1.0444	1.0470	1.0496	1.0522
	1.0697	1.0766	1.0835	1.1903	1.0971	1.1039	1.1107	1.1175	1.1242	1.1310
1.20	1.0548	1.0573	1.0598	1.0624	1.0649	1.0674	1.0698	1.0723	1.0747	1.0772
	1.1377	1.1444	1.1511	1.1578	1.1645	1.1711	1.1777	1.1844	1.1910	1.1976
1.30	1.0797	1.0820	1.0844	1.0867	1.0891	1.0914	1.0938	1.0961	1.0984	1.1007
	1.2042	1.2107	1.2173	1.2238	1.2304	1.2369	1.2434	1.2499	1.2564	1.2629
1.40	1.1030	1.1052	1.1075	1.1097	1.1119	1.1141	1.1163	1.1185	1.1207	1.1229
	1.2693	1.2758	1.2822	1.2886	1.2951	1.3015	1.3079	1.3142	1.3206	1.3270
1.50	1.1250	1.1271	1.1293	1.1314	1.1335	1.1356	1.1376	1.1397	1.1418	1.1438
	1.3333	1.3397	1.3460	1.3523	1.3587	1.3650	1.3713	1.3775	1.3838	1.3901
1.60	1.1458	1.1479	1.1499	1.1519	1.1539	1.1559	1.1578	1.1598	1.1617	1.1637
	1.3963	1.4026	1.4088	1.4151	1.4213	1.4275	1.4337	1.4399	1.4461	1.4523
1.70	1.1656	1.1675	1.1694	1.1713	1.1732	1.1751	1.1770	1.1788	1.1807	1.1825
	1.4585	1.4646	1.4708	1.4769	1.4831	1.4892	1.4954	1.5015	1.5067	1.5137
1.80	1.1844	1.1862	1.1880	1.1898	1.1916	1.1934	1.1952	1.1970	1.1987	1.2005
	1.5198	1.5259	1.5320	1.5381	1.5441	1.5502	1.5562	1.5623	1.5683	1.5744
1.90	1.2022	1.2040	1.2057	1.2074	1.2091	1.2180	1.2125	1.2142	1.2159	1.2176
	1.5804	1.5864	1.5925	1.5985	1.6045	1.6105	1.6165	1.6225	1.6284	1.6344

†$1.00 \leq n_D/n_A \leq 1.99$

Problem 8.3

Design a periodic multilayer with suppressed second and third orders using only two coating materials with $n_L = 1.45$ and $n_H = 2.35$ (Sec. 8.1.4).

Solution. From Table 8.1 we determine $n_B = 1.9544$. From Eq. 8.5 $n_E = n_A n_B/n_C = 1.206$. The desired multilayer is consequently built around the antireflection coating 1.206 |AB| 2.35 with $n_A = 1.45$ and $n_B = 1.9544$ which has zero reflectance at the wavenumber positions $\lambda/\lambda_0 = 2$ and 3 (solid curve in Fig. 8.12). Using Ohmer's equations (Sec. 3.1.4) at the quarter-wave position we obtain the dotted (high-index layers on outside) or the dashed (low-index layers on outside) curves. We observe that there is a good match only at the point of forced equality. Because of its better overall shape we decide to improve the antireflection coating with low indices on the outside matching further. After entering we determine a reflectance of 0.049 percent at $\lambda/\lambda_0 = 2$

TABLE 8.6. Relative Refractive Indices n_A/n_C, n_D/n_C as a Function of n_B/n_C of a Structure (ABCDDCBA) which Generates Suppressed First, Second, Fourth, Fifth, Sixth, and Seventh Orders If Used as a Base Period of a Periodic Multilayer†

n_B/n_C	n_A/n_C n_D/n_C									
	0.00	0.01	0.02	0.03	0.04	0.05	0.06	0.07	0.08	0.09
2.00	1.2192	1.2209	1.2225	1.2242	1.2258	1.2274	1.2290	1.2307	1.2323	1.2339
	1.6404	1.6464	1.6523	1.6583	1.6642	1.6702	1.6761	1.6820	1.6880	1.6939
2.10	1.2354	1.2370	1.2386	1.2402	1.2417	1.2433	1.2448	1.2464	1.2479	1.2494
	1.6998	1.7057	1.7116	1.7175	1.7234	1.7293	1.7352	1.7410	1.7469	1.7528
2.20	1.2510	1.2525	1.2540	1.2555	1.2570	1.2585	1.2599	1.2614	1.2629	1.2643
	1.7587	1.7645	1.7704	1.7762	1.7821	1.7879	1.7937	1.7996	1.8054	1.8112
2.30	1.2658	1.2672	1.2687	1.2701	1.2715	1.2730	1.2744	1.2758	1.2772	1.2786
	1.8170	1.8229	1.8287	1.8345	1.8403	1.8461	1.8519	1.8577	1.8634	1.8692
2.40	1.2800	1.2814	1.2828	1.2841	1.2855	1.2869	1.2882	1.2896	1.2909	1.2923
	1.8750	1.8808	1.8865	1.8923	1.8981	1.9038	1.9096	1.9153	1.9211	1.9268
2.50	1.2936	1.2950	1.2963	1.2976	1.2989	1.3002	1.3015	1.3028	1.3041	1.3054
	1.9325	1.9383	1.9440	1.9497	1.9555	1.9612	1.9669	1.9726	1.9783	1.9840
2.60	1.3067	1.3080	1.3093	1.3105	1.3118	1.3131	1.3143	1.3156	1.3168	1.3181
	1.9897	1.9954	2.0011	2.0068	2.0125	2.0182	2.0239	2.0295	2.0352	2.0409
2.70	1.3193	1.3205	1.3217	1.3230	1.3242	1.3254	1.3266	1.3278	1.3290	1.3302
	2.0466	2.0522	2.0579	2.0635	2.0692	2.0749	2.0805	2.0862	2.0918	2.0974
2.80	1.3341	1.3326	1.3337	1.3349	1.3361	1.3373	1.3384	1.3396	1.3407	1.3419
	2.1031	2.1087	2.1143	2.1200	2.1256	2.1312	2.1368	2.1425	2.1481	2.1537
2.90	1.3430	1.3442	1.3453	1.3464	1.3476	1.3487	1.3498	1.3509	1.3520	1.3531
	2.1593	2.1649	2.1705	2.1761	2.1817	2.1873	2.1929	2.1985	2.2041	2.2097

†$2.00 \leq n_D/n_A \leq 2.99$

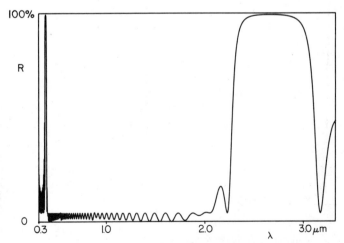

Figure 8.11. Infrared reflector in the 2.5-μm region with good transmittance down through the visual: 1 | (L M1 M2 HH M2 M1 L)^5 | 1.52 with $n_L = 1.46$, $n_{M1} = 1.68$, $n_{M2} = 2.04$, and $n_H = 2.35$ ($\lambda_0 = 0.65$ μm).

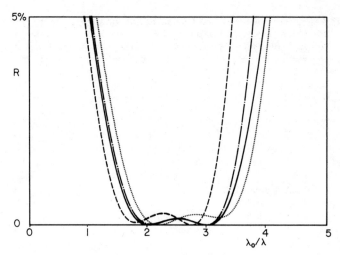

Figure 8.12. Design of an antireflection coating for the suppression of the second and third stopbands of a periodic multilayer: $1.206 \mid 0.4(AB) \mid 2.35$ (solid curve), $1.206 \mid 0.5105A\ 0.1651C\ 0.1105A \mid 2.35$ (dashed curve), $1.206 \mid 0.4A\ 0.1467C\ 0.0964A\ 0.1467C \mid 2.35$ (dotted curve), and $1.206 \mid 0.4494A\ 0.1558C\ 0.0971A \mid 2.35$ (dash-dotted curve) with $n_A = 1.45$, $n_B = 1.9544$, and $n_C = 2.35$.

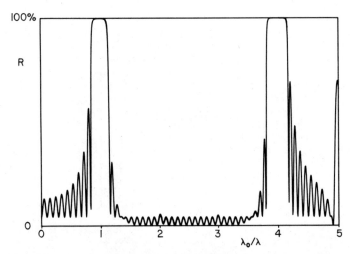

Figure 8.13. Periodic two-material multilayer with suppressed second and third orders: $1 \mid (0.4494L\ 0.1558H\ 0.0971L\ 0.58H\ 0.0971L\ 0.1558H\ 0.4494L)^{10} \mid 1.52$ with $n_L = 1.45$ and $n_H = 2.35$.

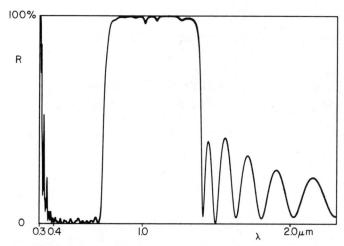

Figure 8.14. Dual-band heat reflector design using essentially two materials: $1 \mid 1.1(\text{L}/2 \text{ H L}/2) \,(\text{L}/2 \text{ H L}/2)^5 \, 1.125(\text{L}/2 \text{ H L}/2)$ $1.374(0.4494\text{L }0.1558\text{H }0.0971\text{L }0.58\text{H }0.0971\text{L }0.1558\text{H }0.4494\text{L})^8$ $0.64\text{L}2 \mid 1.52$ with $n_\text{L} = 1.45$, $n_\text{H} = 2.35$, and $n_{\text{L}2} = 1.38$ ($\lambda_0 = 0.860$ μm).

and 0.006 percent at $\lambda/\lambda_0 = 3$. With refining (Chap. 11) these residual reflectances could be reduced to 0.011 and 0.002 percent. The resulting antireflection coating is the dash-dotted curve of Fig. 8.12. Figure 8.13 shows a periodic multilayer using this structure and Fig. 8.14 shows an application as dual-band heat reflector.

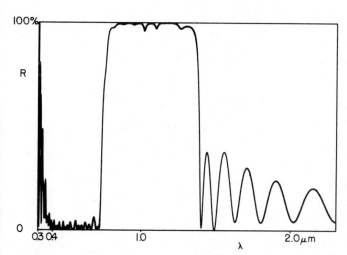

Figure 8.15. Dual-band heat reflector design with higher low-index material: $1 \mid 1.1 \,(\text{L}/2 \text{ H L}/2) \,(\text{L}/2 \text{ H L}/2)^5 \, 1.125(\text{L}/2 \text{ H L}/2) \, 0.55$ $(\text{LMHML})^8 \mid 1.52$ with $n_\text{L} = 1.45$, $n_\text{H} = 2.35$, and $n_\text{M} = 1.9544$ ($\lambda_0 = 0.860$ μm).

Problem 8.4

Redesign the dual-stack heat reflector of Fig. 8.5 using as low index $n_L = 1.45$.

Solution.   Exchanging 1.45 for 1.38 while maintaining the optical layer thicknesses leads to the design of Fig. 8.15. Adding a quarter wave with index 1.38 between the design and the substrate improves the transmittance in the passband (Figs. 8.16 and 6.10).

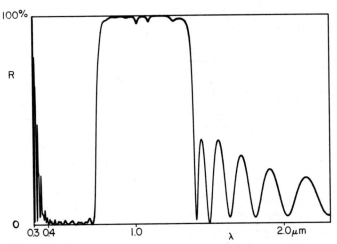

Figure 8.16. The same design as in Fig. 8.15, but now a MgF$_2$-anti-reflection coating was added between the design and the substrate: $1 \mid 1.1(L/2 \text{ H } L/2)$ $(L/2 \text{ H } L/2)^5 \; 1.125(L/2 \text{ H } L/2) \; 0.55(\text{LMHML})^8$ $\mid 0.64\text{L2} \mid 1.52$ with $n_L = 1.45$, $n_H = 2.35$, $n_M = 1.9544$, and $n_{L2} = 1.38$ ($\lambda_0 = 860$ nm).

Edge Filters at Nonnormal Incidence: Thin Film Polarizers and Nonpolarizing Edge Filters

As we discussed before (Sec. 2.10), the reflectance and the transmittance of optical interference films change dramatically when the light incidence is changed from normal to oblique incidence. For edge filters we have an increase in the width for the stopband for p-polarization and a decrease for s-polarization since the spread between the effective high and low refractive indices increases for s-polarization and decreases for p-polarization. The centers of the stopbands remain aligned but shift to shorter wavelengths since the effective optical thicknesses decrease equally in both planes of polarization (Eq. 2.5). Figure 9.1 shows the angle shift of the edge filter design of Fig. 2.7.

Another consequence of Eq. 2.5 is that at nonnormal incidence the optical thicknesses of layers with different refractive indices assume different effective values since α is related to α_0 through Snell's law: $n \sin \alpha = n_0 \sin \alpha_0$. A quarter-wave stack at normal incidence is no longer a quarter-wave stack at nonnormal incidence. We show an important consequence of this fact in Fig. 9.2. The suppressed second order of a quarter-wave stack shows up as a *half-wave hole* at nonnormal light incidence. (Another common cause of the half-wave hole is dispersion, see Chap. 12.)

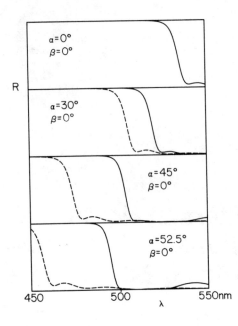

Figure 9.1. Shift of the s- and p-reflectances of the design 1 | (H/2 L H/2)^{12} 0.95(H/2 L H/2)^3 | 1.52 with $n_L = 1.45$ and $n_H = 2.35$ ($\lambda_0 = 450$ nm), with incidence angles $\alpha_0 = 0°, 30°, 45°,$ and $52.5°$. The match angle is $\beta = 0$ in all cases. Reflectance scales are 0 to 100 percent [s-polarization (solid curves) and p-polarization (dashed curves)].

In order to compensate for the loss of optical thickness match at nonnormal light incidence and also for the shift of the stack center to shorter wavelength, all optical thicknesses are divided by cos β where β is the so-called *match angle* (Eq. 2.58). In Fig. 9.3 we show the design of Fig. 9.1 matched for every incidence angle.

For edge filters cemented in a glass cube ($n_0 = n_s > 1.5$) an acceleration of the angle deterioration takes place. At 52.5° light incidence, the stopband of the p-plane of the design of Fig. 9.4 almost completely disappears.

9.1. Thin Film Polarizers

9.1.1. MacNeille Polarizer†

We saw in Fig. 2.6 that the effective index of two different materials at nonnormal incidence and p-polarization can become equal:

$$\frac{n_A}{\cos \phi_A} = \frac{n_B}{\cos \phi_B} \tag{9.1}$$

---

†From MacNeille.[1]

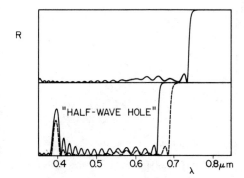

R

"HALF-WAVE HOLE"

0.4   0.5   0.6   0.7   0.8 μm
λ

Figure 9.2. s- and p-reflectances of the visual short wave pass of Fig. 6.10, at $0°$ (top curve) and $45°$ (bottom curve) light incidence: $1 \mid 1.065(L2/2 \ H \ L2/2)^2 (L2/2 \ H \ L2/2)^{11} \ 1.065(L2/2 \ H \ L2/2)^2 \ 1.5L1 \mid 1.52$ with $n_{L1} = 1.38$, $n_{L2} = 1.45$, and $n_H = 2.35$ ($\lambda_0 = 0.86$ μm, $\beta = 0°$); s-polarization (solid curves) and p-polarization (dashed curves). Reflectance scales are 0 to 100 percent.

ϕ_A is, of course, the Brewster angle for two materials with refractive indices n_A and n_B. Snell's law demands

$$n_A \sin \phi_A = n_B \sin \phi_B = n_0 \sin \phi_0 \qquad (9.2)$$

Combining Eqs. 9.1 and 9.2 yields

$$n_0^2 \sin^2 \phi_0 = \frac{n_A^2 \ n_B^2}{n_A^2 + n_B^2} \qquad (9.3)$$

If one thin film material has a much larger refractive index than the other (for $n_A \gg n_B$, $n_H = n_A$ and $n_L = n_B$; for $n_B \gg n_A$, $n_H = n_B$ and $n_L = n_A$), then

$$n_0 \sin \alpha_0 \approx \frac{n_L}{n_H} \qquad (9.4)$$

We can see that Eq. 9.3 is difficult to fulfill. MacNeille[1] used in his patent $n_L = 1.38$ (MgF$_2$), $n_H = 2.4$ (ZnS), $\alpha_0 = 45°$, and $n_0 = 1.69$. For further information on MacNeille polarizers also see Banning[1a]. Figure 9.5 gives an example of the patent.

9.1.2. Edge Polarizers

It is obvious from Figs. 9.1 to 9.4 that between the stopband of p-polarization and s-polarization there are wavelength regions where the reflectance is high for p-polarization and low for s-polarization. By using as incidence angle the Brewster angle of the substrate, the reflectance of the second surface can be minimized. Figure 9.6 gives an example.

The usable zone is only 5 percent wide. There is consequently not much room for the usual errors in thickness, refractive index, and uniformity.

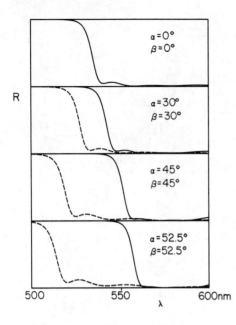

Figure 9.3. s- and p-reflectances of the design $1 \mid (H/2\ L\ H/2)^{12}\ 0.95(H/2\ L\ H/2)^3 \mid 1.52$ with $n_L = 1.45$ and $n_H = 2.35$ ($\lambda_0 = 450$ nm) for the four incidence angles $\alpha_0 = 0°, 30°, 45°,$ and $52.5°$. The optical thicknesses are matched for each incidence angle ($\alpha_0 = \beta$). Reflectance scales are 0 to 100 percent [s-polarization (solid curves) and p-polarization (dashed curves)].

9.2. Nonpolarizing Edge Filters

If we want to design an edge filter for use at nonnormal incidence there is not much difficulty when the incident light is linearly polarized and the filter can be positioned so that the plane of polarization is either

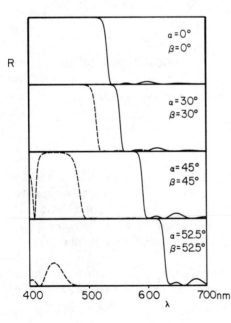

Figure 9.4. s- and p-reflectances of the "immersed" edge filter design $1.52 \mid 0.95(H/2\ L\ H/2)^3\ (H/2\ L\ H/2)^9\ 0.95(H/2\ L\ H/2)^3 \mid 1.52$ with $n_L = 1.45$ and $n_H = 2.35$ ($\lambda_0 = 450$ nm) for various incidence angles $\alpha_0 = 0°, 30°, 45°,$ and $52.5°$ ($\alpha_0 = \beta$) [s-polarization (solid curves) and p-polarization (dashed curves)].

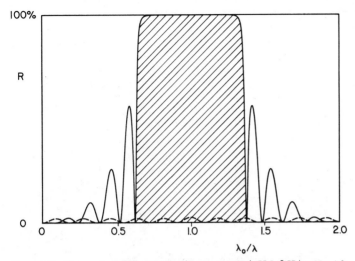

Figure 9.5. Example of a MacNeille polarizer: $1.69 \mid (\text{H L})^5 \text{ H} \mid 1.69$ with $n_L = 1.38$ and $n_H = 2.40$ ($\alpha_0 = \beta = 45°$) [s-polarization (solid curve) and p-polarization (dashed curve)]. As predicted, the multilayer behaves like a thick single film in the p-plane. Performance can be improved with better matching.

vertical or parallel to the substrate (Sec. 2.10). In all other cases great difficulties arise. The sharpness of the edge is no longer limited by the number of layers but by the degree of polarization. Figure 9.7 shows the design of Fig. 2.7 for unpolarized light.

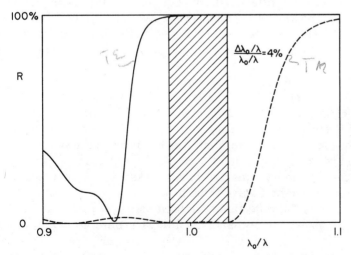

Figure 9.6. Reflectance of the polarizer design $1 \mid 0.735(\text{L}/2 \text{ H LH}/2)^2$ $0.84(\text{L}/2 \text{ H L}/2)^8$ $0.735\,(\text{L}/2 \text{ H L}/2)^2 \mid 1.52$ with $n_L = 1.45$, $n_H = 2.35$, and $\alpha_0 = \beta = 56.7°$ [s-polarization (solid curve) and p-polarization (dashed curve)].

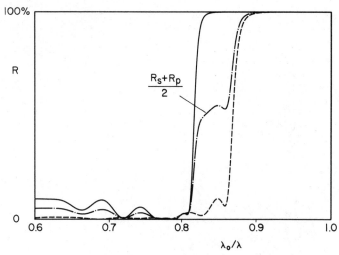

Figure 9.7. Average reflectance $(R_s + R_p)/2$ for the design of Fig. 2.7 $(\alpha_0 = \beta\ 45°)$.

9.2.1. Analog Method of Design†

Figure 9.8 gives the reflectance of a typical single-spacer all-dielectric narrow bandpass filter (Chap. 10) for 45° light incidence. The optical thicknesses are matched for the incidence angle ($\beta = \alpha_0$). The reflectance curves for the two planes of polarization are different, but they are centered at the same λ_0/λ position.

Let us apply the split filter formula (Eq. 2.55) to this filter. LL is the spacer layer and the two quarter-wave stacks n_L | HLHLHLH |1 and n_L | HLHLHLH | 1.52 are Systems A and B. Since the reflectance of a quarter-wave stack around the position $\lambda_0/\lambda = 1$ is almost constant $[T_A T_B/(1 - r)^2 \approx 1$ and $r \approx 1]$ the filter effect is solely generated by changes of $\Theta = \Phi_A + \Phi_B - 2\varphi$. The position of minimum overall reflectance is determined by $\Theta = \pi$. Figure 9.9 shows the difference in phase on reflection (from Eq. 2.37) in the two planes of polarization $\Phi_{Bs} - \Phi_{Bp}$ of System B. Only at $\lambda_0/\lambda = 1$ is the difference zero. This means that the positions of the minimum reflectances in the two planes of polarization coincide only when the minimum reflectances coincide with the centers of the two quarter-wave stacks, or, in other words, when the spacer is one-half-wave thick at λ_0.

Figure 9.10 shows the reflectances in both planes of polarization of a detuned single spacer narrow bandpass filter. The minimum positions are now different.

The extension of the single spacer narrow bandpass filter is the

---

†From Thelen.[2]

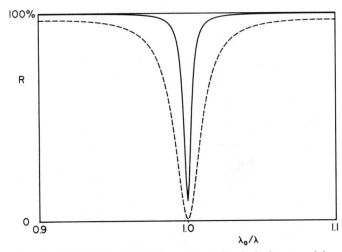

Figure 9.8. Reflectances R_s (solid curve) and R_p (dashed curve) of the narrow bandpass design 1 | HLHLHLH LL HLHLHLH | 1.52 with $n_H = 2.28$, $n_L = 1.45$, and $\alpha_0 = \beta = 45°$.

multiple spacer bandpass filter (Chap. 10). The construction is medium | (matching layers) (half-wave spacer) (reflecting stack) (half-wave spacer) (reflecting stack) (half-wave spacer) · · · (reflecting stack) (half-wave spacer) (matching layers) | substrate. If we now detune the half-wave spacers we should be able to influence the shift of the reflectance characteristics in the one plane of polarization relative to the other. If

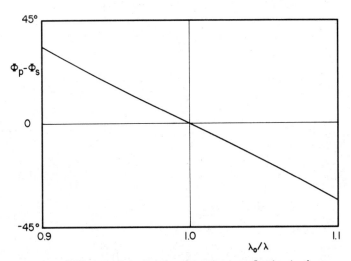

Figure 9.9. Difference between the phases upon reflection in the two planes of polarization for the configuration 1.45 | HLHLHLH | 1.52 with $n_H = 2.28$, $n_L = 1.45$, and $\alpha_0 = \beta = 45°$.

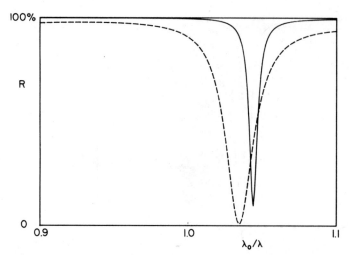

Figure 9.10. Reflectances R_s (solid curve) and R_p (dashed curve) of the design of Fig. 9.8, but now the cavity is detuned: 1 | HLHLHLH 1.8L HLHLHLH | 1.52 with $n_H = 2.28$, $n_L = 1.45$, and $\alpha_0 = \beta\ 45°$.

we manage to bring the edges in the two planes of polarization into alignment, we have generated a nonpolarizing edge filter, as shown in Fig. 9.11. The fact that the range of low reflectance next to the edge is only of limited width and that the steepness of the edges in the two

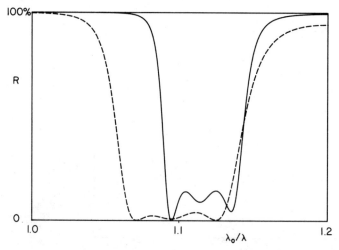

Figure 9.11. Reflectances R_s (solid curve) and R_p (dashed curve) of the triple cavity filter 1 | HLH 1.54L HLHLHLH 3.295L HLHLHLH 1.54L HLH | 1.52 with $n_H = 2.28$, $n_L = 1.45$, and $\alpha_0 = \beta = 45°$. The amount of detuning of the outer cavities was adjusted for alignment of the edges and the inner cavities for ripple in the passband.

planes of polarization is different can be a problem, but is of little consequence in many applications.

We arrive at the following design procedure for a nonpolarizing edge filter:

1. Select a reflecting stack as the basic building block of a bandpass filter. Remember that the higher the number of layers the narrower the usable high-transmittance range.

2. Detune spacers by trial and error to align the edges in the two planes of polarization.

3. Determine the matching layers that minimize the secondary reflectance bands.

4. Reduce the secondary reflectance bands further by refining (Chap. 11).

Figure 9.12 gives a complete short-wavelength-pass design and Fig. 9.13 a long-wavelength-pass design. Incident medium is vacuum or air. Both designs went through an extensive refining procedure. The deviations from the original optical thicknesses, as given in the captions, are given in Table 9.1.

For the case of an edge filter cemented inside a glass cube, Fig. 9.14 gives a long-wavelength-pass design and Fig. 9.15 a short-wavelength-pass design. These designs were not refined.

The reflecting stack in the basic structure need not be a quarter-

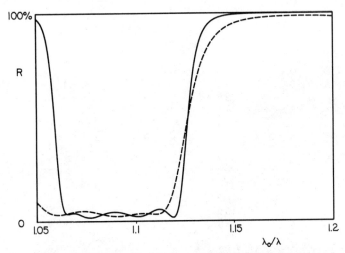

Figure 9.12. Reflectances R_s (solid curve) and R_p (dashed curve) of the short-wavelength-pass filter with the starting design $1 \mid H\ 0.8L\ (0.8L\ HLHLHLH\ 0.8L)^4\ 0.8L\ HLH \mid 1.52$ with $n_H = 2.28$, $n_L = 1.45$, and $\alpha_0 = \beta\ 45°$. Final thicknesses are given in Table 9.1.

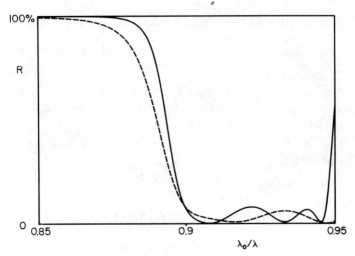

Figure 9.13. Reflectances R_s (solid curve) and R_p (dashed curve) of the long-wavelength-pass filter with the starting design $1 \mid H\ 1.2L\ (1.2L\ HLHLHLH\ 1.2L)^4\ 1.2L\ HLH \mid 1.52$ with $n_H = 2.28$, $n_L = 1.45$, and $\alpha_0 = \beta = 45°$. Final thicknesses are given in Table 9.1.

TABLE 9.1. Deviation in Percentages from the Starting Optical Thicknesses of the Designs in Figs. 9.12 and 9.13

Layer number	Design of Fig. 9.12, %	Design of Fig. 9.13, %
1	+36.9	−66.1%
2	−21.5	+37.8
3	+17.6	−50.5
4	+14.3	−25.4
5	+4.0	+32.4
6	−2.0	−22.1
7	−2.5	−12.1
8	−2.9	+31.9
9	−4.0	0.0
10	+2.4	−3.2
11–25	No Changes	
26	+5.4	+6.7
27	−30.3	−31.3
28	+23.3	−9.7
29	+23.3	+18.2
30	−25.3	+34.6
31	−15.2	+13.5
32	−51.1	−18.4
33	+49.3	−37.8
34	−8.1	+18.4
35	−0.9	−37.2
36	+23.2	−7.3
37	+22.1	+63.6

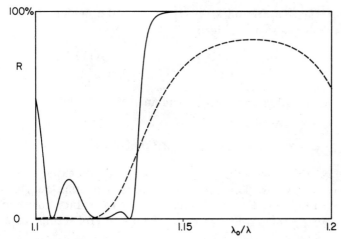

Figure 9.14. Reflectances R_s (solid curve) and R_p (dashed curve) of the long-wavelength-pass filter cemented inside a cube 1.52 | HLH 0.8L (0.8L HLHLHLH 0.8L)^4 0.8L HLH | 1.52 with $n_H = 2.28$, $n_L = 1.45$, and $\alpha_0 = \beta = 45°$.

wave stack. As has been shown by Thelen,[2a] a so-called half/quarter stack (. . . HHLHHLHHLHHL . . . or . . . LLHLLHLLHLLH . . .) can also be used. We must remember, though, that the centers of the stopbands for these designs do not occur at $\lambda_0/\lambda = 1, 3, 5, \ldots$ as in the case of quarter-wave stacks, but at the positions $\lambda_0/\lambda = {}^2/_3, {}^4/_3, {}^8/_3, {}^{12}/_3, \ldots$. This means that we must apply a factor of ${}^2/_3$ to all the thicknesses to

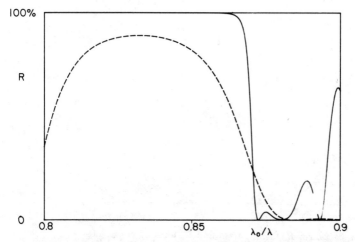

Figure 9.15. Reflectances R_s (solid curve) and R_p (dashed curve) of the short-wavelength-pass filter cemented inside a cube 1.52 | HLH 1.25L (1.25L HLHLHLH 1.25L)^4 1.25L HLH | 1.52 with $n_H = 2.28$, $n_L = 1.45$, and $\alpha_0 = \beta = 45°$.

make the filter effective around $\lambda_0/\lambda = 1$ (we use 0.6, which is close enough to establish the starting design):

(Half spacer reflecting stack, half spacer)
$$= (1.2H\ 0.6L\ 1.2H \cdots) \text{ or } (1.2L\ 0.6H\ 1.2L \cdots) \qquad (9.6)$$

In order to align the edges, we detune the spacers and find, by trial and error,

$$Z = 1.3H\ 0.6L\ 1.2H\ 0.6L\ 1.2H\ 0.6L\ 1.3H \qquad (9.7)$$

We are surprised that the detuning is smaller than we experienced with a quarter-wave stack. The explanation is that for non-quarter-wave stacks, the difference $\Phi_s - \Phi_p$ is not zero for $\lambda_0/\lambda = 1$ (or, in other words, we are getting some inherent help through the reflecting stack in lining up the edges).

Figure 9.16 gives the resulting short-wavelength-pass design. The method of shifted periods (Chap. 6) was used to reduce pass band ripple. This design is of particular interest since it has a wider passband region than the previous designs.

9.2.2. Analytical Method of Design†

The previously introduced analog method (Sec. 9.2.1) of designing non-polarizing edge filters is limited to modifying the spacer layers of a

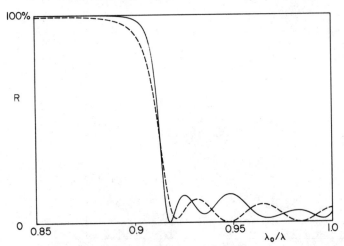

Figure 9.16. Reflectances R_s (solid curve) and R_p (dashed curve) of the short-wavelength-pass filter $1\ |\ H\ 1.02Z\ Z^4\ HLH\ |\ 1.52$ with $Z = 1.3H\ 0.6L\ 1.2H\ 0.6L\ 1.2H\ 0.6L\ 1.3H$ $(n_H = 2.28,\ n_L = 1.45,\ \alpha_0 = \beta = 45°)$.

---

†From Thelen.[3]

bandpass filter. Yet there are more general solutions to the problem. Let us consider the most general structure of a symmetric period of a multilayer edge filter:

$$a_1A_1\ a_2A_2 \cdots a_{m-1}A_{m-1}\ a_mA_m \mid a_mA_m\ a_{m-1}A_{m-1} \cdots a_2A_2\ a_1A_1 \quad (9.8)$$

where A_x is a layer with refractive index n_x and quarter-wave thickness λ_0 and $\Sigma a_x = 1$ for $x = 1$ to m, $0 < a_x < 1$. The symmetry of the structure is accentuated by the vertical line.

Structure 9.8 can be considered consisting of two parts: the half-structure

$$a_1A_1\ a_2A_2 \cdots a_{m-1}A_{m-1}\ a_mA_m \quad (9.9)$$

and the same structure arranged in reverse order.

The characteristic matrix of the half-structure is

$$\mathbf{Q} = \mathbf{B}_1\ \mathbf{B}_2 \cdots\ \mathbf{B}_{m-1}\ \mathbf{B}_m$$

$\mathbf{B}_x$ stands for the layer a_xA_x. Setting

$$\mathbf{B} = \mathbf{B}_1\ \mathbf{B}_2\ \cdots\ \mathbf{B}_{m-2}$$

we arrive at $\mathbf{Q} = \mathbf{B}\ \mathbf{B}_{m-1}\ \mathbf{B}_m$ or

$$\mathbf{Q} = \mathbf{B} \begin{bmatrix} \cos\phi_{m-1} & \dfrac{i\sin\phi_{m-1}}{n_{m-1}} \\ in_{m-1}\sin\phi_{m-1} & \cos\phi_{m-1} \end{bmatrix} \begin{bmatrix} \cos\phi_m & \dfrac{i\sin\phi_m}{n_m} \\ in_m\sin\phi_m & \cos\phi_m \end{bmatrix} \quad (9.10)$$

This yields for the element Q_{11} of the matrix $\mathbf{Q}$:

$$Q_{11} = B_{11}\left(\cos\phi_{m-1}\cos\phi_m - \frac{n_m\sin\phi_{m-1}\sin\phi_m}{n_{m-1}}\right)$$
$$- B_{12}\left(n_{m-1}\sin\phi_{m-1}\cos\phi_m + \cos\phi_{m-1}\sin\phi_m\right) \quad (9.11)$$

and similar expressions for Q_{12}, Q_{21}, and Q_{22}.

With Eq. 2.27 we can write for the characteristic matrix of the complete period Eq. 9.8

$$\mathbf{M} = \begin{bmatrix} M_{11} & iM_{12} \\ iM_{21} & M_{22} \end{bmatrix} = \begin{bmatrix} Q_{11} & iQ_{12} \\ iQ_{21} & Q_{22} \end{bmatrix}\begin{bmatrix} Q_{22} & iQ_{12} \\ iQ_{21} & Q_{11} \end{bmatrix}$$
$$= \begin{bmatrix} Q_{11}Q_{22} - Q_{12}Q_{21} & 2iQ_{11}Q_{12} \\ 2iQ_{22}Q_{21} & -Q_{12}Q_{21} + Q_{11}Q_{22} \end{bmatrix} \quad (9.12)$$

At the edges of the stopband, either M_{12} or M_{21} equals zero (Sec. 6.1.1).

In terms of the elements of the half-structure (Eq. 9.9) this translates, with Eq. 9.12, into

$$Q_{11} \quad \text{or} \quad Q_{12} \quad \text{or} \quad Q_{21} \quad \text{or} \quad Q_{22} = 0 \qquad (9.13)$$

Setting $Q_{11} = 0$ yields, from Eq. 9.11,

$$\tan \phi_{m-1} = \frac{B_{11}/B_{12} - n_m \tan \phi_m}{n_{m-1} + B_{11}n_m \tan \phi_m / B_{12}n_{m-1}} \qquad (9.14)$$

At nonnormal incidence, the effective refractive indices in the two planes of polarization are different (Eq. 2.16), but the phase thicknesses ϕ_x are not. We consequently can use Eq. 9.14 to define aligned edges as follows:

$$\left[\frac{B_{11}/B_{12} - n_m \tan \phi_m}{n_{m-1} + B_{11}n_m \tan \phi_m / B_{12}n_{m-1}} \right]_p$$

$$- \left[\frac{B_{11}/B_{12} - n_m \tan \phi_m}{n_{m-1} + B_{11}n_m \tan \phi_m / B_{12}n_{m-1}} \right]_s = 0 \qquad (9.15)$$

which is a quadratic equation for $\tan \phi_m$.

Similar quadratic equations for $\tan \phi_m$ can be derived by setting the other matrix elements of $\mathbf{Q}$ equal to zero (Thelen[3]). These four quadratic equations have up to eight solutions for a periodic multilayer with the basic structure 9.8 and $m - 2$ given phase thicknesses ϕ_1, $\phi_2, \ldots, \phi_{m-1}$. The various solutions define nonpolarizing edges at different sides of the stopbands (long- or short-wavelength sides) and for different orders of the stopbands. Some solutions are redundant.

Two-material five-element structures. The simplest combination of structure 9.8 yielding nonpolarizing edges is a two-material five-element structure:

$$a_1 A_1 \, a_2 A_2 \, (a_3 A_1 \, a_3 A_1) \, a_2 A_2 \, a_1 A_1 \qquad (9.16)$$

For this case

$$B_{11} = \cos \phi_1 \qquad B_{12} = \frac{\sin \phi_1}{n_1}$$

$$B_{21} = n_1 \sin \phi_1 \qquad B_{22} = \cos \phi_1$$

With given refractive indices n_1 and n_2, ϕ_1 can be changed at will. How do we find the optimum solution? One way is to calculate ϕ_2 and ϕ_3 as a function of ϕ_1 and maximize the deviation of $\Sigma = \phi_1 + \phi_2 + \phi_3$ from the next multiple of 90°. This deviation is an indication of the width of the stopband. Another way is to calculate the reflectance for the combination of phase angles found. Neglecting matching influences, the higher the reflectance the better the design. Figure 9.17 gives the result of such an optimization for the two coating materials $n_1 = 3.5$ and $n_2 = 1.45$.

Fine tuning the edges.   Figure 9.17 exposes a flaw in the design procedure used so far. The reflectance curves in the two planes of polarization are aligned at the transition point from oscillation to a steady decrease of the reflectance as a function of the number of periods. Much more desirable would be the 50 percent point as an alignment point. This can be accomplished by subjecting the solutions of Eq. 9.15 to an additional adjustment procedure. Equation 9.15 is of the type

$$F_p(n_1, n_2, \ldots, n_m; \phi_1, \phi_2, \ldots, \phi_{m-1}, \phi_m)$$
$$- F_s(n_1, n_2, \ldots, n_m; \phi_1, \phi_2, \ldots, \phi_{m-1}, \phi_m) = 0$$

To allow fine tuning we make the following modifications:

$$\tan^{-1}[F_p(n_1, n_2, \ldots, n_m; \phi_1, \phi_2, \ldots, \phi_{m-1}, \phi_m)] - (1 + \delta)$$
$$\tan^{-1}[F_s(n_1, n_2, \ldots, n_m; \phi_1, \phi_2, \ldots, \phi_{m-1}, \phi_m)] = 0 \qquad (9.17)$$

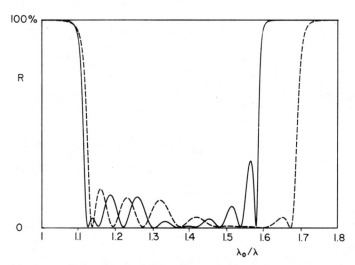

Figure 9.17.   Reflectance in the two planes of polarization of the design 1 | 0.7S1 (0.5161H 0.2876L 0.3725H 0.2876L 0.5161H)^2 (0.4213H 0.4905L 0.1763H 0.4905L 0.4213H)^5 (0.5161H 0.2876L 0.3725H 0.2876L 0.5161H)^2 0.7S2 | 1.52 with $n_H = 3.5$, $n_L = 1.45$, $n_{S1} = 1.72$, $n_{S2} = 2.13$, and $\alpha_0 = \beta = 45°$ as a function of the relative wavenumber λ_0/λ. S1 and S2 are temporary matching layers which later are replaced by synthesized H-L combinations. The two double periods around the five times repeated core period were designed using the method of similar equivalent layers described in Chap. 7.

where δ is a small number. Unlike Eq. 9.15, Eq. 9.17 does not allow closed solutions. But with a simple iterative procedure solutions can be readily found.

Figure 9.18 gives an expanded view around the edge of the design of Fig. 9.17. Figure 9.19 gives a new design with better alignment of the edges. This design is the result of applying Eq. 9.17 with $\delta = -1\%$.

The final design with good edge alignment and improved matching is then given in Fig. 9.20.

Nonpolarizing edge filters using nonsymmetric periods. As we discussed before (Sec. 8.1), the edge of a periodic multilayer, composed of periods with the characteristic matrix elements M_{11}, M_{12}, M_{21}, and M_{22}, is defined by

$$M_{11} + M_{22} = +2 \quad \text{or} \quad -2$$

Let us now assume that the period structure is

$$a_1A_1\, a_2A_2 \cdots a_{m-2}A_{m-2}\, a_{m-1}A_{m-1}\, a_mA_m$$

The first $m - 2$ layers are combined and described by characteristic matrix **B**. $M_{11} + M_{22}$ can then be put into the following form:

$M_{11} + M_{22}$

$$= \left[(B_{11} + B_{22})\cos \phi_m - \left(B_{12}n_m + \frac{B_{21}}{n_m} \right) \sin \phi_m \right] \cos \phi_{m-1}$$

$$- \left[\left(B_{12}n_{m-1} + \frac{B_{21}}{n_{m-1}} \right) \cos \phi_m \right.$$

$$\left. + \left(\frac{B_{11}n_m}{n_{m-1}} + \frac{B_{22}n_{m-1}}{n_m} \right) \sin \phi_m \right] \sin \phi_{m-1}$$

or $\qquad M_{11} + M_{22} = u \cos \phi_{m-1} + v \sin \phi_{m-1} = \pm 2$

This equation can be solved for ϕ_{m-1}

$$\phi_{m-1} = \cos^{-1}\left(\frac{\pm 2}{\sqrt{u^2 + v^2}} \right) + \tan^{-1}\frac{v}{u} = F(\phi_m) \qquad (9.18)$$

which leads to a similar equation as before for ϕ_m (Eq. 9.17)

$$F_p\,(\phi_m) - (1 + \delta)F_s\,(\phi_m) = 0 \qquad (9.19)$$

As an example we take the four-element structure aA bB cA dB as period, $n_A = n_L = 1.45$ and $n_B = n_H = 2.35$. **B** applies to aA bB.

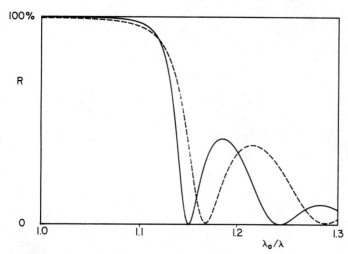

Figure 9.18. Reflectance in the two planes of polarization of the design
1 | 0.7S1(0.4213H 0.4905L 0.1763H 0.4905L 0.4213H)^6 0.7S2 | 1.52
with n_H = 3.5, n_L = 1.45, n_{S1} = 1.72, n_{S2} = 2.13, and α_0 = β = 45°.

Using the same optimization procedure as before (page 190), but now
Eq. 9.19 instead of Eq. 9.15, we arrive at the structure
| 0.6445L 0.6445H 0.3908L 0.3202H | providing a nonpolarizing edge
for the short-wave side of the first stopband. Figure 9.21 gives the
resulting design. In Fig. 3.8 we improved the transmittance near the
edge using the equivalent layer concept for nonsymmetric periods.

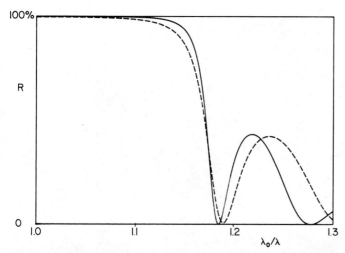

Figure 9.19. Reflectance in the two planes of polarization of the design
1 | 0.7S1 (0.4116H 0.5249L 0.1271H 0.5249L 0.4116H)^6 0.7S2 | 1.52
with n_H = 3.5, n_L = 1.45, n_{S1} = 1.72, n_{S2} = 2.13, and α_0 = β = 45°.

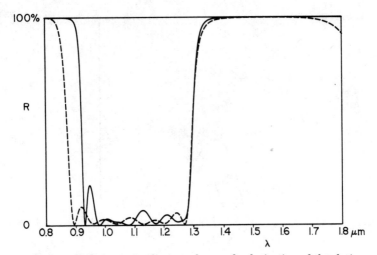

Figure 9.20. Reflectance in the two planes of polarization of the design
1 | (0.1678H 0.5613L 0.1678H) (0.9143H 0.1867L 0.7880H
0.1867L 0.9143H) (0.7320H 0.4648L 0.4680H 0.4648L
0.7320H) (0.6153H 0.7847L 0.1900H 0.7847L 0.6153H)^4 (0.7320H
0.4648L 0.4680H 0.4648L 0.7320H) (0.9143H 0.1867L 0.7880H
0.1867L 0.9143H) (0.2752H 0.3467L 0.2752H) | 1.52   with $n_H = 3.5$,
$n_L = 1.45$, and $\alpha_0 = \beta = 45°$ ($\lambda_0 = 1$ μm).

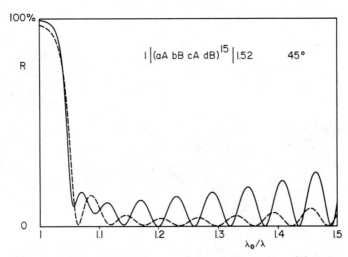

Figure 9.21. Reflectance in the two planes of polarization of the nonpolarizing   edge   filter   design   with   nonsymmetric   periods:
1 | (0.6445L 0.6445H 0.3908L 0.3202H)^{15} | 1.52   with $n_L = 1.45$,
$n_H = 2.35$, and $\alpha_0 = \beta = 45°$.

9.3. Problems and Solutions

Problem 9.1

With Fig. 9.2 we introduced the *half-wave hole* as an indication of the fact that an equal thickness (quarter-wave) stack at normal incidence is not an equal thickness stack at oblique incidence. Now, let us assume that a half-wave hole appears at normal incidence, indicating that the stack is not properly matched, how can one determine whether the high-index layer is thicker than the low-index layer or vice versa?

Solution. The optical thickness of a high-index layer decreases less rapidly with increasing incidence angle than the optical thickness of the low-index layer. Consequently, when the half-wave hole becomes more severe when the incidence angle is increased, the low-index layer must be heavier, when it becomes less severe, the high-index layer must be heavier.

Problem 9.2

Design a short-wave-pass nonpolarizing edge filter to be cemented into a quartz ($n_0 = n_s = 1.45$) cube and to be used at $20°$ light incidence using the detuned band filter technique. Coating materials should be silicon dioxide ($n_L = 1.45$) and titanium dioxide ($n_H = 2.2$).

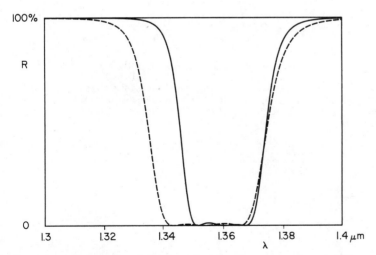

Figure 9.22. Reflectance of the nonpolarizing long-wave-pass edge filter 1.45 | H L H L H 2.24L H L H L H L H L H L H 4.34L H L H L H L H L H L H 2.24 L H L H L H | 1.45 with $n_L = 1.45$, $n_H = 2.2$, and $\alpha = \beta = 20°$ ($\lambda_0 = 1.3$ μm).

Solution. We offer two solutions: one based on nonsymmetric periods (Fig. 9.21) and one based on the analog method (Fig. 9.22).

Problem 9.3

Design a nonpolarizing short-wave-pass edge filter with silicon ($n_H = 3.5$) and silicon dioxide ($n_L = 1.45$) as coating materials. The design is again to be cemented in a quartz cube ($n_0 = n_s = 1.45$) and to be used at 20° light incidence. Use the analytical method of design.

Solution. The resulting design is given in Fig. 9.23.

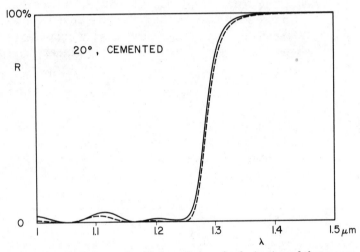

Figure 9.23. Reflectance in the two planes of polarization of the nonpolarizing edge filter design 1.45 | (0.2752H 0.3468L 0.2752H) (0.9143H 0.1867L 0.788H 0.1867L 0.9143H) (0.732H 0.4648L 0.468H 0.4648L 0.732H) (0.6153H 0.7847L 0.18H 0.7847L 0.6153H)⁴ (0.732H 0.4648L 0.468H 0.4648L 0.732H) (0.9143H 0.1867L 0.788H 0.1867L 0.9143H) (0.2752H 0.3468L 0.2752H) | 1.45 with $n_L = 1.45$, $n_H = 3.5$, and $\alpha_0 = \beta = 20°$ ($\lambda_0 = 1$ μm).

Narrow Bandpass Filters

A narrow bandpass filter has high transmittance in a narrow wavelength region (λ_1 to λ_2) and high rejection (low transmittance rsp. high reflectance) in all other wavelength regions ($\lambda < \lambda_1$ and $\lambda > \lambda_2$). The transition from the rejection regions to the passband should be as rapid as possible (*square* bandpasses).

Narrow bandpass filters consist in general of two parts:

1. A design which generates the actual narrow bandpass characteristic (transition from low to high transmittance band, a high transmittance band, and the transition from high to low transmittance)

2. Blocking filters which provide rejection in wavelength regions where, due to their periodic nature, the narrow bandpass designs have high transmittance zones

In this section we will limit ourselves to the design of actual narrow bandpasses. Blocking filters are generally edge filters and are discussed in Chap. 6 and especially in Sec. 6.3.

10.1. Design by Elimination of Half-Wave Layers

The earliest approach to the construction of narrow bandpasses (Geffcken[1]) was based on simulating a Fabry-Perot etalon (Fabry and Perot[2]) with the structure

Glass | semitransparent silver film | half-wave dielectric spacer film |
semitransparent silver film | glass

In the same patent, Geffcken[1] also proposed filters of the type

Glass | semitransparent silver film | half-wave dielectric spacer film |
semitransparent silver film | half-wave dielectric spacer film |
semitransparent silver film | glass

The step to replace the semitransparent silver films with a stack of
dielectric layers came soon (Geffcken,[3] Polster,[4] Turner[5]), leading to
designs of the type

$$\text{HLH} \cdots \text{L HH LHL} \cdots \text{L HH LHL} \cdots \text{H} \qquad (10.1)$$

Structures of the type (Eq. 10.1) lend themselves to design by elimination of half-wave layers (Sec. 3.4). Since only half-wave and quarter-
wave layers are used, every elimination of a half-wave layer leads to
the generation of another half-wave layer until only quarter-wave layers are left. The number of quarter-wave layers left then determines
the maximum transmittance of the narrow bandpass filter design. For
example, the filter

$$1 \,|\, \text{LHLHLHL HH LHLHLH} \,|\, 4 \qquad (10.2)$$

can be reduced to

$$
\begin{aligned}
&1 \,|\, \text{LHLHLH LL HLHLH} \,|\, 4 \\
&1 \,|\, \text{LHLHL HH LHLH} \,|\, 4 \\
&1 \,|\, \text{LHLH LL HLH} \,|\, 4 \\
&1 \,|\, \text{LHL HH LH} \,|\, 4 \\
&1 \,|\, \text{LH LL H} \,|\, 4 \\
&1 \,|\, \text{L HH} \,|\, 4 \\
&1 \,|\, \text{L} \,|\, 4
\end{aligned}
\qquad (10.3)
$$

Consequently, at the maximum, the transmittance of this design is the
same as that of a single low-index layer. Figure 10.1 gives the reflectance of designs (Eqs. 10.2 and 10.3) with $n_H = 4.0$ and $n_L = 1.8$.

10.2. Design by Using Split Filter Analysis[†]

Smith[5a] demonstrated his design method with the following double
half-wave filter:

$$1 \,|\, \text{HL HH LHLHL HH L1H} \,|\, 1 \qquad (10.4)$$

[†]See Sec. 3.4 and Smith.[5a]

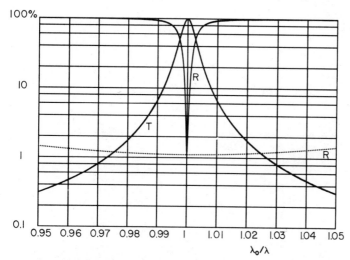

Figure 10.1. Reflectance and transmittance of the single spacer narrow-band infrared filter design 1 | LHLHLHL HH LHLHLH | 4 (solid curves) with $n_L = 1.8$ and $n_H = 4$. Also shown is the reflectance of a single layer with $n_L = 1.8$ (1 | L | 4, dotted curve).

For given n_H and n_L, what is the optimum value for n_{L1}?

Following Smith,[5a] we split the filter at the second half wave and generate the two systems:

System A $\qquad\qquad$ n_H | LHLHL HH LH | 1 $\qquad\qquad$ (10.5)

and

System B $\qquad\qquad$ n_H | L1H | 1 $\qquad\qquad$ (10.6)

For $n_L = 1.35$ and $n_H = 4.0$, Fig. 10.2 gives the reflectance of System A and System B for three values of n_{L1}: 1.6, 1.83, and 2.0. If the reflectances of Systems A and B intersect twice we have two maxima and if they intersect once we have one maximum. The transmittance at the maximum is 100 percent when Eq. 3.45 is fulfilled also. If the reflectances of Systems A and B do not intersect at all the transmittance has a single maximum below 100 percent. Figure 10.3 shows the corresponding composite transmittance characteristics.

10.3   Design with Equivalent Layers†

Let us return to the double-wave design of the previous section:

$$n_0 \mid \text{HL HH LHLHL HH L1H} \mid n_S$$

---

†From Thelen.[5b]

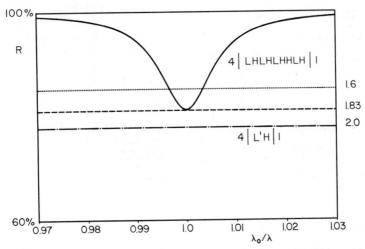

Figure 10.2. Reflectance of the configuration $4\,|\,$LHLHL HH LH$\,|\,1$ with $n_L = 1.35$ and $n_H = 4$ (System A, solid curve) and $4\,|\,$L1H$\,|\,1$ (System B) with $n_H = 4$ and $n_{L1} = 1.6$ (dotted curve), 1.83 (dashed curve), and 2.0 (dash-dotted curve).

We can break this design down into the following three equivalent layers:

Equivalent layer I   HLH

Equivalent layer II   HLHLHLH   (10.7)

Equivalent layer III   HL1H

The transmittance is high when

$$N_I = \sqrt{n_0 N_{II}} \qquad \Gamma_I = (2m + 1)90° \qquad m = 0, 1, 3, \ldots$$
$$N_{III} = \sqrt{n_S N_{II}} \qquad \Gamma_{III} = (2m + 1)90° \qquad m = 0, 1, 3, \ldots$$

(10.8)

Interpretation 10.7 opens up a wide variety of possible narrow bandpass designs. Equivalent layer II can be repeated several times without upsetting the matching relations (Eq. 10.8). In this way, the transition from passband to stopbands can be sharpened and the rejection level improved without altering the bandwidth and the peak transmittance.

Figure 10.4 shows a design where equivalent layer II is repeated twice. Equivalent layer III uses a lower index for the low-index film in order to provide the better match to air. Figure 10.5 gives a design which uses a 2 : 1 optical thickness ratio configuration as equivalent layer II. A filter of this type is less angle-sensitive than a comparable design with equally thick layers (Sec. 2.10).

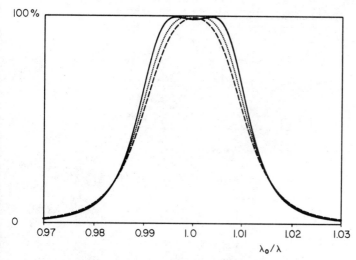

Figure 10.3. Transmittance of the narrow-band filter design
1 | HL HH LHLHL HH L1H | 1 with $n_L = 1.35$, $n_H = 4$, and $n_{L1} = 1.6$
(solid curve), 1.83 (dotted curve), and 2.0 (dashed curve).

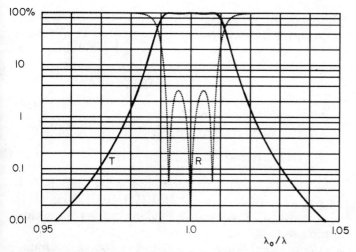

Figure 10.4. Transmittance of the narrow bandpass filter
1 | (HL1H) (HLHLHLHLH)² (HLHL) | 4 with $n_H = 4$, $n_L = 1.8$, and
$n_{L1} = 1.6$. Reflectance is also shown as the dotted curve.

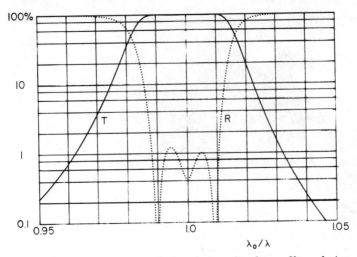

Figure 10.5. Transmittance of the narrow bandpass filter design
1 | (HL1H) (6H/5 3L/5 6H/5 3L/5 6H/5 3L/5 6H/5 3L/5 6H/5)^2 (HLH)| 1.7
with $n_H = 4$, $n_L = 1.8$, and $n_{L1} = 2.2$. Reflectance is also shown as the
dotted curve.

The model can be expanded to make predictions about bandwidth
and angle shift (Thelen[5b]). With the high speed and ease of modern
computers these calculations are rarely very helpful in design practice.

10.3.1. Mixed half-wave and full-wave cavities

Generally, the design of narrow bandpass filters becomes difficult when
the main equivalent layer II is repeated more than twice. This is shown
in Fig. 10.6. Additional flexibility can be generated by including some
half-wave layers into the core period. For example, the following filters
are, except for the multiplicity of the half waves, identical:

$$HLH \ (HLHLHLH)^4 \ HLH \qquad (10.9)$$

$$HL \ (HHLHLHLHH)^4 \ LH \qquad (10.10)$$

$$HL \ (HHLHLHLHHLHLHLHH)^2 \ LH \qquad (10.11)$$

The spacer sequence is HH, HH, HH, HH, HH for the first, HH, HHHH,
HHHH, HHHH, HH for the second, and HH, HH, HHHH, HH, HH for
the third design. Figure 10.7 gives the equivalent indices of the three
variations assuming $n_L = 1.8$ and $n_H = 4$. Figure 10.8 shows a five-
cavity design with equivalent layer (Eq. 10.10). The equivalent indices
at $\lambda_0/\lambda = 1$ are $N_{HHLHLHLHH} = 0.3645$, $N_{L1HL1} = 0.64$, and $N_{L2HL2} =
0.525$ with $n_H = 4$, $n_L = 1.8$, $n_{L1} = 1.6$, and $n_L = 1.45$.

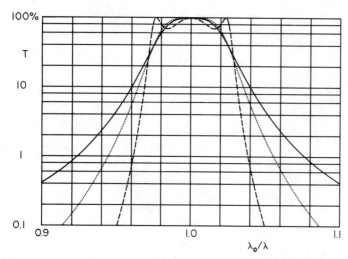

Figure 10.6. Transmittance of the narrow bandpass designs 1 | HLH HLHLHLH HLHL | 1.34 (solid curve), 1 | LHLH (HLHLHLH)^2 HLHL | 1.34 (dotted curve), and 1 | HLH (HLHLHLH)^3 HLHL | 1.34 (dashed curve) with n_L = 1.8 and n_H = 4.

Figure 10.9 shows a five-cavity design using equivalent layer (Eq. 10.11). Now $N_{\text{HHLHLHLHHHLHLHLHH}}$ = 0.418 and N_{L1HL1} = 0.64 with n_H = 4, n_L = 1.8, and n_{L1} = 1.6.

Another advantage of the mixed half-wave full-wave designs is that they allow additional flexibility in the adaptation to a given bandwidth.

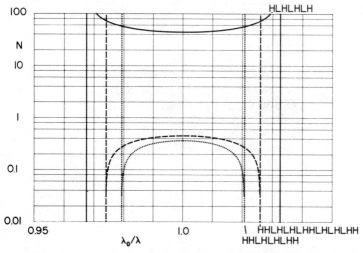

Figure 10.7. Equivalent indices of the structures H L H L H L H, H H L H L H L H L H H, and H H L H L H L H L H H H L H L H H with n_L = 1.8 and n_H = 4.

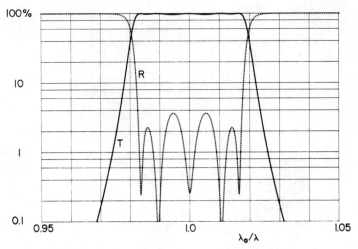

Figure 10.8. Transmittance (solid curve) and reflectance (dotted curve) of the narrow bandpass design 1 | L1HL1 (HHLHLHLHH)^4 L2HL2 | 1.34 with $n_H = 4$, $n_L = 1.8$, $n_{L1} = 1.45$, and $n_{L2} = 1.6$.

10.3.2. Composite spacers

The location and quality of narrow bandpass interference filters are much more sensitive to errors in the spacer layers than to errors in any other layer. When the optical thickness of the spacer exceeds one half wave this sensitivity is further increased. One way to overcome

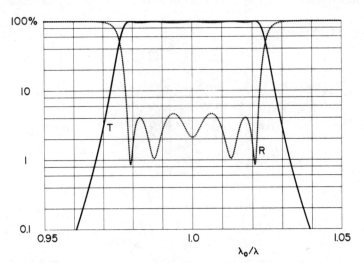

Figure 10.9. Transmittance (solid curve) and reflectance (dotted curve) of the narrow bandpass design 1 | L 1H L 1 (H H L H L H L H H L H L H L H H)^2 L 1 H L 1 | 1.34 with $n_H = 4$, $n_L = 1.8$, and $n_{L1} = 1.6$.

this problem is to use a composite spacer, e.g., HHLLHH instead of HHHHHH. In Figure 10.10 we show two designs based on the core equivalent layer LHHLHLHLHLHLHHL with $n = 2.35$ and $n = 1.45$.

10.3.3. Design with similar equivalent layers

Another way to overcome the problem of poor transmittance in the passband is to use similar equivalent layers (Sec. 7.1). The narrow bandpass design of Fig. 10.11 uses, instead of four equal equivalent layers in the core, two inner ones and two similar (meaning the same structure but different indices) outer ones. The outer similar equivalent layers provide, together with equivalent layers I and III, a better match to the medium and the substrate.

10.3.4. Designs with more than two materials in the core[†]

We slowly increased the number of refractive indices which we used in our narrow bandpass designs from two to four. At first we limited the additional indices to the matching layers but in the last subsection we used them closer to the core. A systematic utilization of three refractive indices to accommodate a large variety of given bandwidths

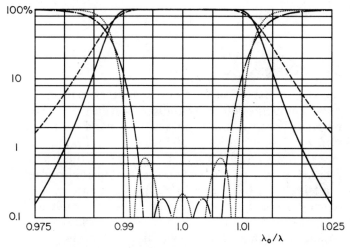

Figure   10.10 Characteristics   of   the   designs   1.52 | LHLH (LHHLHLHLHLHLHHL)^p HLHL | 1.52 with $n_H = 2.35$ and $n_L = 1.45$. Transmittance for $p = 2$ is dashed curve and for $p = 3$ is solid curve. Reflectance for $p = 2$ is dash-dotted curve and for $p = 3$ dotted curve.

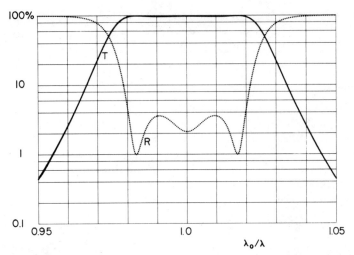

Figure 10.11 Transmittance (solid curve) and reflectance (dotted curve) of the narrow bandpass design 1 | MHMH (HMHMHMH) (HLHLHLH)^2 (HMHMHMH) HMHM | 1.34 with $n_H = 4$, $n_M = 2.2$, and $n_L = 1.8$.

was proposed by Jacobs.[6] The reader is referred to the original paper for details. The given designs are to be taken as demonstrations and not as evidence what the method can ultimately accomplish. There is one word of caution, though: the author could not verify the extension to three materials of the concept of the "effective index ratio ρ" to the prediction of the bandwidth.

10.4. Equal Ripple Designs

From tables published by Levy[6a] or with our computer program (Problem 3.5) we can determine the refractive indices for a desired prototype filter corresponding to a desired number of layers D, bandwidth W, and rejection level $(n_S/n_0)_{ERAR}$. As an example, we show in Fig. 10.12 the refractive index as a function of the optical thickness of the 1 percent equal ripple filter of Table 3.10. As we can see, the desired refractive indices are far outside the range one can realize in practice.

We offer two methods to synthesize these extremely high or low refractive indices: (1) finding an equivalent layer whose equivalent index matches the desired index, or (2) looking at the prototype filter as a series of cavities and reflectors. The reflectors are generated by quarter-wave stacks (Baumeister[6b]).

Both methods maintain the amplitudes of the interface reflections but not the phases since the half-wave layers are replaced by composite structures with much larger thickness. The result is that the trans-

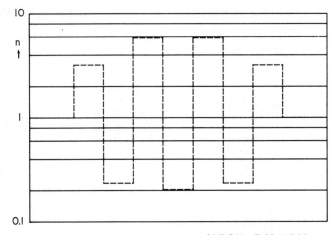

── OPTICAL THICKNESS

Figure 10.12. Refractive index as a function of the optical thickness of the 1 percent equal ripple filter of Table 3.10.

mittance level, ripple, and off-band rejection are maintained but the bandwidth is not. This is a severe limitation and puts this method on par with other methods.

10.4.1. Equal ripple designs with equivalent layers

In Prob. 3.4 we derived a formula for the equivalent index of a sequence of quarter-wave-thick layers (Eq. 3.54).

$$N_{\text{quarter-wave stack}} = \frac{n_1 n_3 n_5 \cdots}{n_2 n_4 n_6 \cdots} \qquad (3.54)$$

Obviously, very large and very low values for N can be generated this way and it appears that sequences of quarter-wave-thick layers should be well suited to synthesize the layers of equal ripple filters. They have to be used twice, though, since their equivalent thickness is an odd multiple of a quarter wave while the equal ripple filter requires (multiples of) a half wave.

Let us demonstrate the method on a series of filters using three coating materials ($n_L = 1.45$, $n_M = 1.95$, and $n_H = 4.3$) and equal massive media ($n_0 = n_S = 1$).

In Table 10.1 we list the equivalent indices of all possible symmetric seven-layer combinations of the selected materials. Two features of Table 10.1 are important.

TABLE 10.1 Equivalent Index of All Symmetric Seven-Layer Combinations with $n_L = 1.45$, $n_M = 1.95$, and $n_H = 4.3$

LLLLLLL	1.45	MLLLLLM	2.622	HLLLLLH	12.752
LLLMLLL	1.078	MLLMLLM	1.95	HLLMLLH	9.482
LLLHLLL	0.489	MLLHLLM	0.884	HLLHLLH	4.3
LLMLMLL	2.622	MLMLMLM	4.743	HLMLMLH	23.062
LLMMMLL	1.95	MLMMMLM	3.527	HLMMMLH	17.149
LLMHMLL	0.884	MLMHMLM	1.599	HLMHMLH	7.777
LLHLHLL	12.752	MLHLHLM	23.062	HLHLHLH	112.142
LLHMHLL	9.482	MLHMHLM	17.149	HLHMHLH	83.388
LLHHHLL	4.3	MLHHHLM	7.777	HLHHHLH	37.815
LMLLLML	0.802	MMLLLMM	1.45	HMLLLMH	7.051
LMLMLML	0.596	MMLMLMM	1.078	HMLMLMH	5.243
LMLHLML	0.270	MMLHLMM	0.489	HMLHLMH	2.378
LMMLMML	1.45	MMMLMMM	2.622	HMMLMMH	12.752
LMMMMML	1.078	MMMMMMM	1.95	HMMMMMH	9.482
LMMHMML	0.489	MMMHMMM	0.884	HMMHMMH	4.3
LMHLHML	7.051	MMHLHMM	12.752	HMHLHMH	62.006
LMHMHML	5.243	MMHMHMM	9.482	HMHMHMH	46.107
LMHHHML	2.378	MMHHHMM	4.3	HMHHHMH	20.909
LHLLLHL	0.165	MHLLLHM	0.298	HHLLLHH	1.45
LHLMLHL	0.123	MHLMLHM	0.222	HHLMLHH	1.078
LHLHLHL	0.056	MHLHLHM	0.101	HHLHLHH	0.489
LHMLMHL	0.298	MHMLMHM	0.539	HHMLMHH	2.622
LHMMMHL	0.222	MHMMMHM	0.401	HHMMMHH	1.95
LHMHMHL	0.101	MHMHMHM	0.182	HHMHMHH	0.884
LHHLHHL	1.45	MHHLHHM	2.622	HHHLHHH	12.752
LHHMHHL	1.078	MHHMHHM	1.95	HHHMHHH	9.482
LHHHHHL	0.489	MHHHHHM	0.884	HHHHHHH	4.3

1. It contains the equivalent indices of all possible symmetric combinations below seven layers. Also, according to Eq. 3.54 equal indices on odd and even places can be cancelled, e.g., $N_{MMHLHMM} = N_{HLH}$, $N_{MHMMMHM} = N_{MHMHM}$.

2. Layers on odd/even places can be exchanged as long as the odd/even character is maintained, e.g., $N_{HHLLLHH} = N_{LHHLHHL}$, $N_{MLHHHLM} = N_{HLMHMLH}$.

As a basis for our synthesization, let us now use an equal ripple filter with three layers, a fractional bandwidth $W_{ERAR} = 0.66$, and $(n_S/n_0)_{ERAR} = 1000$. The subscript ERAR stands for equal ripple antireflection coating. An $(n_S/n_0)_{ERAR} = 1000$ corresponds to a $T_{min} \approx 4/(n_S/n_0)_{ERAR} = 0.4\%$. We determine by interpolation from Levy's tables (Levy[6a]) or with our computer program (Problem 3.5) the following prototype 1 | HH LL HH | 1 with $n_H = 4.34$ and $n_L = 0.595$. Since 4.34 is close enough to the index of the high-index element available for the equivalent layers we can take it directly. From Table 10.1

we take N_{LMLMLML} = 0.596. With these synthesizations we arrive at the final design:

$$1.0 \mid \text{HH} \, (\text{LMLMLML} \,)^2 \, \text{HH} \mid 1.0 \qquad (10.12)$$

The transmittance of the design (Eq. 10.12) is shown in Fig. 10.13, dashed curve. We are surprised to find a much higher T_{min} than expected. The reason is the difference in thickness between the middle layer and the outer layers. By increasing the thickness of the outer layers in Fig. 10.13, solid and dotted curves, T_{min} can be lowered.

For our second example we select as equal ripple filter prototype a five-layer combination with W_{ERAR} = 0.5, and $(n_S/n_0)_{\text{ERAR}}$ = 100,000 (T_{min} = 4 × 10^{-3}%). As before, we obtain for our prototype 1 | MM LL HH LL MM | 1 with n_M = 2.58, n_L = 0.304, and n_H = 4.39. From Table 10.1 we select N_{MLM} = N_{LLMLMLL} = 2.622, N_{MHLHM} = N_{MHLLLHM} = 0.298, and N_{MLMLMLM} = 4.743. The result is shown in Fig. 10.14.

For our third example we use again a five-layer prototype but this time with narrower bandwidth W_{ERAR} = 0.4 and higher off-band rejection $(n_S/n_0)_{\text{ERAR}}$ = 1,000,000 (T_{min} = 4 × 10^{-4}%). The prototype is 1 | MM LL HH LL MM | 1 with the indices n_M = 3.185, n_L = 0.238, and n_H = 5.586. From Table 10.1 N_{MLM} = N_{MMMLMMM} = 2.622, N_{LHMHL} = N_{LHMMMHL} = 0.222, and N_{LMHMHML} = 5.243. Figure 10.15 shows the result.

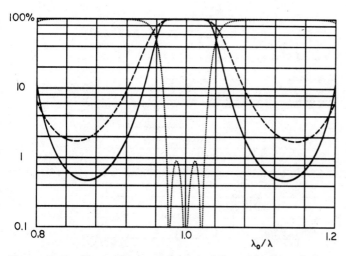

Figure 10.13. Transmittance characteristics of the designs 1 | (HH)^p (LMLMLML)^2 (HH)^p | 1 with n_L = 1.45, n_M = 1.95, and n_H = 4.3 for p = 1 (dashed curve) and p = 2 (solid curve transmittance and dotted curve reflectance).

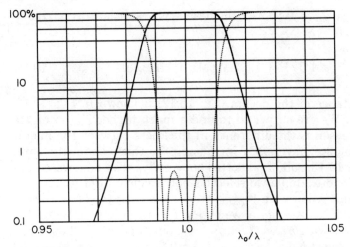

Figure 10.14. Transmittance (solid curve) and reflectance (dotted curve) of the design $1 \mid (MLM)^2 (MHLHM)^2 (MLMLMLM)^2 (MHLHM)^2 (MLM)^2 \mid 1$ with $n_L = 1.45$, $n_M = 1.95$, and $n_H = 4.3$.

Equal ripple bandpasses with nonsymmetric equivalents. Let us reexamine Prob. 3.4 and especially matrix Eq. 3.53. We note that this matrix can be interpreted as a matrix of a single layer independently of whether it represents a symmetric or nonsymmetric combination of layers. The only requirement is that the number of layers must be an odd number. We use Eq. 3.54 to calculate N for all possible symmetric

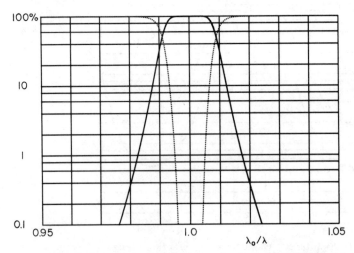

Figure 10.15. Transmittance (solid curve) and reflectance (dotted curve) of the design $1 \mid (MLM)^2 (LHMHL)^2 (LMHMHML)^2 (LHMHL)^2 (MLM)^2 \mid 1$ with $n_L = 1.45$, $n_M = 1.95$, and $n_H = 4.3$.

and nonsymmetric five-layer combinations of the three-material L, M, and H with $n_L = 1.45$, $n_M = 1.95$, and $n_H = 4.3$ (table not shown). For the third example of Sec. 10.4 ($n_M = 3.185$, $n_L = 0.238$, and $n_H = 5.586$) we can now select $N_{LLLMH} = 3.197$, $N_{LHMHL} = 0.222$, and $N_{MHHLH} = 5.586$. The result is shown in Fig. 10.16. It is interesting to note that these designs are invariant to using the nonsymmetric combinations in a forward or reverse direction: (LLLMH) or (HMLLL). Nonsymmetric combinations of this type could have been used in previously discussed design methods as well (Secs. 10.3.1–10.3.4).

Prediction of actual bandwidth.   In our synthesization we have been using equivalent layers with much larger thicknesses as prescribed by the equal ripple prototype. But even when we allow for these much larger thicknesses simple predictions of the bandwidth are not possible since the equivalent thicknesses can considerably deviate from the sum of the optical thicknesses of the individual layers.

In Fig. 10.17 we show a bandpass filter which was synthesized from a prototype [five layers, $W_{ERAR} = 0.28$, and $(n_S/n_0)_{ERAR} = 3 \times 10^7$] as described before: 1 | MM LL HH LL MM | 1 with $n_M = 4.32$, $n_L = 0.164$, and $n_H = 7.92$ from Levy's tables or computer program and 1 | HH (LHLHL)^2 (LMHLHML)^2 (LHLHL)^2 HH | 1 with the help of Table 10.1, compared to the modified prototype filter 1 | MM 10L 14H 10L MM | 1 (the multiplicity of the layers corresponds to the sum of the individual layers of the respective equivalent layer). The bandwidth still differs

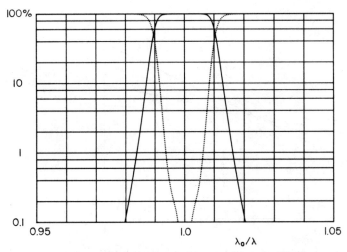

Figure 10.16. Transmittance (solid curve) and reflectance (dotted curve) of the design 1 | (LLLMH)^2 (LHMHL)^2 (MHHLH)^2 (LHMHL)^2 (LLLMH)^2 | 1 with $n_L = 1.45$, $n_M = 1.95$, and $n_H = 4.3$.

212   Chapter 10

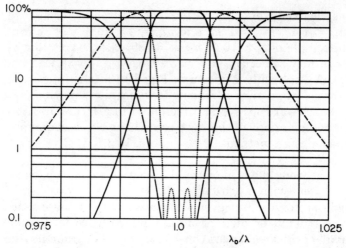

Figure 10.17. Transmittance (solid and dashed curves) and reflectance (dotted and dash-dotted curves) of the designs $1 \,|\, HH$ $(LHLHL)^2$ $(LMHLHML)^2$ $(LHLHL)^2$ $HH \,|\, 1$ with $n_L = 1.45$, $n_M = 1.95$, and $n_H = 4.3$ (solid and dotted curves) and $1 \,|\, MM\ 10L\ 14H\ 10L\ MM \,|\, 1$ with $n_M = 4.32$, $n_L = 0.164$, and $n_H = 7.92$ (dash and dash-dotted curves).

by a factor of two. The explanation can be read from Fig. 10.18 which shows a much more rapid spectral change of the equivalent thickness of the used equivalent layers than the sum of the individual optical thicknesses.

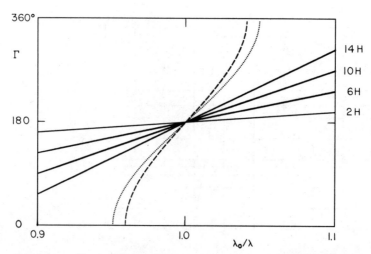

Figure 10.18. Equivalent thicknesses of the sequences 2H, 6H, 10H, and 14H (solid curves), $(LHLHL)^2$ (dotted curve), and $(LMHLHML)^2$ (dashed curve) with $n_L = 1.45$, $n_M = 1.95$, and $n_H = 4.3$.

10.4.2. Equal ripple design with interleaved reflectors

Instead of using equivalent layers to synthesize the individual layers of the prototype we can establish a sequence of half-wave spacers interleaved with a series of dielectric reflectors. The reflectances of these reflectors have to be the same as the corresponding Fresnel reflectances between the layers of the prototype (Baumeister[6b]).

Let us stay with our coating materials n_L = 1.45, n_M = 1.95, and n_H = 4.3 available for synthesization. We now have to prepare tables of the reflectances of all possible L M H combinations for $(n_0/n_S)_{reflector}$ values corresponding to all possible incident medium/spacer, spacer/spacer, and spacer/substrate combinations: 1/1.45, 1/1.95, 1/4.3, 1.45/1.45, 1.45/1.95, 1.45/4.3, 1.95/1.95, 1.95/4.3, and 4.3/4.3 (assuming $n_{0,NBP\ filter}$ = $n_{S,NBP\ filter}$ = 1). As an example, Table 10.2 gives the reflectances of all L M H combinations for one $(n_0/n_S)_{reflector}$ case.

We then select an equal ripple filter prototype and calculate the interface reflectances. For example, for the prototype 1 | MLHLM | 1

TABLE 10.2 Reflectances of the Sequences 4.3 | AB CD | 4.3 with A, B, C, and D Assuming All Possible Combinations of L, M, H with n_L = 1.45, n_M = 1.95, and n_H = 4.3 ($\lambda = \lambda_0$)

LLLL	0.00	MLLL	8.29	HLLL	63.33
LLLM	8.29	MLLM	0.00	HLLM	43.41
LLLH	63.33	MLLH	43.41	HLLH	0.00
LLML	8.29	MLML	28.27	HLML	77.74
LLMM	0.00	MLMM	8.29	HLMM	63.33
LLMH	43.41	MLMH	20.96	HLMH	8.29
LLHM	43.41	MLHM	63.33	HLHM	91.07
LLHH	0.00	MLHH	8.29	HLHH	63.33
LMLL	8.29	MMLL	0.00	HMLL	43.41
LMLM	28.27	MMLM	8.29	HMLM	20.96
LMLH	77.74	MMLH	63.33	HMLH	8.29
LMML	0.00	MMML	8.29	HMML	63.33
LMMM	8.29	MMMM	0.00	HMMM	43.41
LMMH	63.33	MMMH	43.41	HMMH	0.00
LMHL	43.41	MMHL	63.33	HMHL	91.07
LMHM	20.96	MMHM	43.41	HMHM	84.43
LMHH	8.29	MMHH	0.00	HMHH	43.41
LHLL	63.33	MHLL	43.41	HHLL	0.00
LHLM	77.74	MHLM	63.33	HHLM	8.29
LHLH	94.96	MHLH	91.07	HHLH	63.33
LHML	43.41	MHML	20.96	HHML	8.29
LHMM	63.33	MHMM	43.41	HHMM	0.00
LHMH	91.07	MHMH	84.43	HHMH	43.41
LHHL	0.00	MHHL	8.29	HHHL	63.33
LHHM	8.29	MHHM	0.00	HHHM	43.41
LHHH	63.33	MHHH	43.41	HHHH	0.00

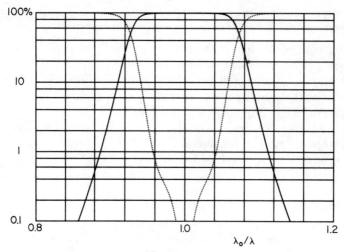

Figure 10.19. Transmittance (solid curve) and reflectance (dotted curve) of the design 1 | MH LL HM HH MHM HH MH M HH MH LL HM | 1 with $n_L = 1.45$, $n_M = 1.95$, and $n_H = 4.3$.

with $n_M = 2.86$, $n_L = 0.163$, and $n_H = 7.32$ [$W_{ERAR} = 0.18$ and $(n_S/n_0)_{ERAR} = 5 \times 10^6$] we calculate the following interface reflectances:

$$R_{0,1} = 23.22 \qquad R_{1,2} = 79.54 \qquad R_{2,3} = 91.44$$

$$R_{2,4} = 91.44 \qquad R_{4,5} = 79.54 \qquad R_{5,6} = 23.22$$

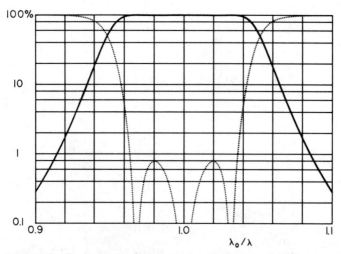

Figure 10.20. Transmittance (solid curve) and reflectance (dotted curve) of the design 1 | LM HH LML HH LHM HH MHL HH LML HH ML | 1 with $n_L = 1.45$, $n_M = 1.95$, and $n_H = 4.3$.

TABLE 10.3 Reflectances of Some Simple Combinations to Match Interference Reflectances. $n_L = 1.45$, $n_M = 1.95$, $n_H = 4.3$.

Reflectance	Design of Fig. 10.19, %	Design of Fig. 10.20, %
$R_{0,1}$	$R_{1|MH|1.45} = 29.22$	$R_{1|LM|4.3} = 16.63$
$R_{1,2}$	$R_{1.45|HM|4.3} = 75.74$	$R_{4.3|LML|4.3} = 77.74$
$R_{2,3}$	$R_{4.3|MHM|4.3} = 84.43$	$R_{4.3|LHM|4.3} = 91.07$

From Table 10.2 and similar tables described above but not repeated here we select the synthesizations displayed in Figs. 10.19 and 10.20 and found in Table 10.3. For both designs $R_{3,4} = R_{3,2}$, $R_{4,5} = R_{2,1}$, and $R_{5,6} = R_{1,0}$. For additional examples see Baumeister.[66].

10.5. Problems and Solutions

Problem 10.1

Adapt the design of Fig. 10.3 from $n_0 = n_S = 1$ to $n_0 = 1$ and $n_S = 1.52$.

Solution. System A (Eq. 10.5) is not affected by the change of n_S from 1 to 1.52 but System B (Eq. 10.6) is. It now becomes $n_H | L1H | 1.52$. In Fig. 10.21 we show the reflectance of System B with lower indices

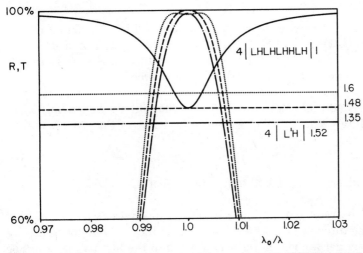

Figure 10.21. Reflectance of the configurations 4 | LHLHLHHLH | 1 (solid curve) and 4 | L1H | 1.52 with $n_{L1} = 1.6$ (dotted curve), $n_{L1} = 1.48$ (dashed curve), and $n_{L1} = 1.35$ (dash-dotted curve). The remaining three unlabeled curves are the transmittance curves of the complete filters 1 | HL HH LHLHL HH L1H | 1.52 with different n_{L1}. The values for n_{L1} and curve patterns of 4 | L1H | 1.52 correspond ($n_H = 4$ and $n_L = 1.8$).

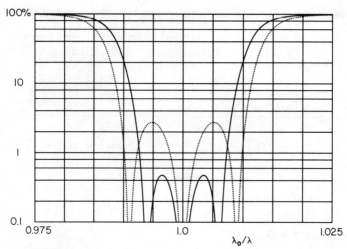

Figure   10.22. Reflectance   of   the   designs   1.45 | HLHLHL (LHLHLHLHLHLHL)² LHLHLH | 1.45 (dotted curve) and 1.45 | HLHLH (LLHLHLHLHLHLHLL)² HLHLH | 1.45 (solid curve) with n_L = 1.45 and n_H = 2.2.

n_{L1} together with the unchanged System A and the three resulting transmittance curves of the finished filters.

Problem 10.2

The reflectance (dotted curve) of the filter 1.45 | HLHLHL (LHLHLHLHLHLHL)² LHLHLH | 1.45 with n_L = 1.45 and n_H = 2.2 is shown in Fig. 10.22. How can the reflectance in the passband be reduced and the bandwidth be slightly narrowed?

Solution.   By adding a half-wave layer in the center a mixed half-wave full-wave filter is generated. See the solid curve of Fig. 10.22.

Problem 10.3

Adapt the filter of Fig. 10.20 to a substrate of n_S = 1.52.

Solution.   The required reflectance for the interface is $R_{5,6}$ = $R_{0,1}$ = 23.22%. The interface 4.2 | 1.52 has a reflectance of 22.82 percent. We consequently get a good match when we drop the last two layers. See Fig. 10.23. The fact that the filter characteristic is determined by the interface reflectance alone and does not directly depend on the refractive indices of the massive media and the cavities is a great advantage of Baumeister's method.

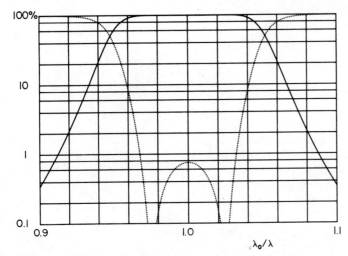

Figure 10.23. Transmittance (solid curve) and reflectance (dotted curve) of the filter 1 | LM HH LML HH LHM HH MHL HH LML HH | 1.52 with $n_\mathrm{L} = 1.45$, $n_\mathrm{M} = 1.95$, and $n_\mathrm{H} = 4.3$.

Computer Refining

In the previous chapters we have learned to find a group of real thin films (thin films which can be deposited in practice) whose reflectance conforms more or less to a given specification. But in order to reduce the complexity we had to use theoretical models which imposed severe restrictions (equal optical thickness, periodicity, symmetry, etc.).

In this chapter we take a different approach. We assume we have found a reasonably good approximation to a problem and we are looking for a way to improve it. We can now lift all the restrictions and ask for the next optimum. We have to make sure, though, that our steps in search of the next optimum are small since we will have to assume linear relationships.

It is obvious that this method only leads us to the local mountain peak and not Mount Everest. For this we would have to use *global* strategies and other high powered mathematical methods which carry the multilayer further and further away from the models which we used to design it and which allowed us to understand its internal functioning. We stick to variations of the original method of refining introduced by Baumeister and Stone.[1] Methods beyond the purpose of this book can be found in Dobrowolski,[2] Bloom,[3] and Lidell.[4]

The transmittance of a group of thin films on a given substrate and with light incidence from a certain medium is a function of the optical thicknesses $QWOT_i$ and the refractive indices n_i of the individual films:

$$T = T(QWOT_1, n_1; QWOT_2, n_2; \ldots; QWOT_m, n_m) \qquad (11.1)$$

By partial differentiation with respect to the $QWOT_i$ we can relate changes of the optical thicknesses to changes in the transmittance:

$$\Delta T \approx \frac{\partial T}{\partial QWOT_1} \Delta QWOT_1 + \frac{\partial T}{\partial QWOT_2} \Delta QWOT_2$$
$$+ \ldots + \frac{\partial}{\partial QWOT_m} \Delta QWOT_m \qquad (11.2)$$

where T, ΔT, and the partial derivatives are, of course, wavelength dependent.

Now we would like to change the transmittance at k wavelength points by a prescribed amount $(\Delta T)_j$. From Eq. 11.2

$$\lambda = \lambda_1 \qquad \Delta T_1 \approx \left[\frac{\partial T}{\partial QWOT_1}\right]_1 \Delta QWOT_1 + \left[\frac{\partial T}{\partial QWOT_2}\right]_1 \Delta QWOT_2$$
$$+ \cdots + \left[\frac{\partial T}{\partial QWOT_m}\right]_1 \Delta QWOT_m$$

$$\lambda = \lambda_2 \qquad \Delta T_2 \approx \left[\frac{\partial T}{\partial QWOT_1}\right]_2 \Delta QWOT_1 + \left[\frac{\partial T}{\partial QWOT_2}\right]_2 \Delta QWOT_2$$
$$+ \cdots + \left[\frac{\partial T}{\partial QWOT_m}\right]_2 \Delta QWOT_m$$

$$\cdots\cdots\cdots\cdots\cdots\cdots\cdots\cdots\cdots\cdots\cdots\cdots\cdots\cdots\cdots\cdots\cdots\cdots\cdots$$

$$\lambda = \lambda_k \qquad \Delta T_k \approx \left[\frac{\partial T}{\partial QWOT_1}\right]_k \Delta QWOT_1 + \left[\frac{\partial T}{\partial QWOT_2}\right]_k \Delta QWOT_2$$
$$+ \cdots + \left[\frac{\partial T}{\partial QWOT_m}\right]_k \Delta QWOT_m$$

or in matrix notation

$$\overrightarrow{\Delta T} = \left[\frac{\partial T}{\partial QWOT}\right] \overrightarrow{\Delta QWOT} \qquad (11.3)$$

where $\overrightarrow{\Delta T}$ is a k-dimensional vector, $\overrightarrow{\Delta QWOT}$ an m-dimensional vector, and

$$\left[\frac{\partial T}{\partial QWOT}\right] \qquad (11.4)$$

an m by k matrix.

If we could solve Eq. 11.3 for $\overrightarrow{\Delta\text{QWOT}}$

$$\overrightarrow{\Delta\text{QWOT}} = \left[\frac{\partial T}{\partial\text{QWOT}}\right]^{-1} \overrightarrow{\Delta T} \tag{11.5}$$

we would have the desired solution. Unfortunately, Eq. 11.5 has a unique solution only when $m = k$ (when the number of layers available for change equals the number of wavelength positions where transmittance changes are desired). This restriction imposes an unacceptable limitation on flexibility.

When $m > k$, Eq. 11.5 is underdetermined and we can impose additional conditions. Baumeister[1] proposed that the thickness of each layer available for change should be altered only a minimum amount. In practice, this has proven to be a very good method because it appears to increase the range of linearity of Eq. 11.2.

When $m < k$, Eq. 11.5 is overdetermined. We now have to look for a solution where the difference between the requested transmittance changes and the possible changes is minimized (Rosen and Eldert[5]). In practice, this is the more difficult method since it appears to decrease the range of linearity of Eq. 11.2 and often requires variable dampening.

11.1. Vertical Refining†

When the number of layers available for change is larger than the number of wavelengths positions where transmittance changes are desired ($m > k$), matrix (11.4) has more rows than columns; its shape is vertical. For this reason we call this version of refining *vertical refining*.

For the additional condition of minimizing the thickness variations we introduce the merit function

$$\Psi = \sum_{i=1}^{m} \Delta\text{QWOT}_i^2 + \sum_{j=1}^{k} \Lambda_j \left(\sum_{i=1}^{m} \left[\frac{\partial T}{\partial\text{QWOT}_i}\right]_j \Delta\text{QWOT}_i - \Delta T_j\right) \tag{11.6}$$

and require its partial derivatives with respect to the thicknesses to be zero

$$\frac{\partial\Psi}{\partial\text{QWOT}_i} = 0 \qquad \text{for } i = 1, 2, 3, \ldots, m \tag{11.7}$$

---

†Or $m > k$.

The Λ_j are Lagrange multipliers.[6] Carrying out the partial differentiations, we obtain, from Eq. 11.6,

$$2\overrightarrow{\Delta\text{QWOT}} + \left[\frac{\partial T}{\partial \text{QWOT}}\right]^{\text{T}} \vec{\Lambda} = 0 \qquad (11.8)$$

Multiplying Eq. 11.8 with matrix 11.4 and using Eq. 11.5 leads to

$$2\overrightarrow{\Delta T} + \left[\frac{\partial T}{\partial \text{QWOT}}\right]\left[\frac{\partial T}{\partial \text{QWOT}}\right]^{\text{T}} \vec{\Lambda} = 0 \qquad (11.9)$$

Solving Eq. 11.9 for $\vec{\Lambda}$ and inserting the resulting expression for $\vec{\Lambda}$ into Eq. 11.8 we obtain as the final solution

$$\overrightarrow{\Delta\text{QWOT}} = \left[\frac{\partial T}{\partial \text{QWOT}}\right]^{\text{T}} \left[\left[\frac{\partial T}{\partial \text{QWOT}}\right]\left[\frac{\partial T}{\partial \text{QWOT}}\right]^{\text{T}}\right]^{-1} \overrightarrow{\Delta T} \qquad (11.10)$$

Within the range of linearity of Eq. 11.2, Eq. 11.10 delivers exact solutions. The thickness variations are minimized using the method of least squares.

11.2. Horizontal Refining†

Matrix 11.4 now has a horizontal shape. We call this version *horizontal refining*.

Because of the underdetermination we can no longer expect exact solutions. We can, though, minimize the difference between the requested and the possible transmittance changes (Rosen and Eldert[5]). Using again the method of least squares (but this time instead of and not in addition to Eq. 11.5) we have to minimize the merit function

$$\Psi = \sum_{j=1}^{k} \left(\sum_{i=1}^{m} \left[\frac{\partial T}{\partial \text{QWOT}_i}\right]_j \Delta\text{QWOT}_i - \Delta T_j\right)^2 \qquad (11.11)$$

After partial differentiation with respect to the thicknesses and setting the derivatives to zero we obtain

$$2\left[\frac{\partial T}{\partial \text{QWOT}}\right]^{\text{T}} \left[\left[\frac{\partial T}{\partial \text{QWOT}}\right]\overrightarrow{\Delta\text{QWOT}} - \overrightarrow{\Delta T}\right] = 0 \qquad (11.12)$$

or

$$\overrightarrow{\Delta\text{QWOT}} = \left[\left[\frac{\partial T}{\partial \text{QWOT}}\right]^{\text{T}}\left[\frac{\partial T}{\partial \text{QWOT}}\right]\right]^{-1}\left[\frac{\partial T}{\partial \text{QWOT}}\right]^{\text{T}} \overrightarrow{\Delta T} \qquad (11.13)$$

---

†Or $m < k$.

11.3. Computation of the Partial Derivatives

Following Baumeister[6] we can compute the partial derivatives of Eq. 11.4 by differentiating Eq. 2.40:

$$\frac{\partial T}{\partial \mathrm{QWOT}_i} = \frac{T^2}{2} \left(\frac{n_0}{n_S} M_{11} \frac{\partial M_{11}}{\partial \mathrm{QWOT}_i} + \frac{n_S}{n_0} M_{22} \frac{\partial M_{22}}{\partial \mathrm{QWOT}_i} \right.$$

$$\left. + n_0 n_S M_{12} \frac{\partial M_{12}}{\partial \mathrm{QWOT}_i} + \frac{1}{n_0 n_S} M_{21} \frac{\partial M_{21}}{\partial \mathrm{QWOT}_i} \right) \quad (11.14)$$

and determining the partial derivatives of the matrix elements by substituting for the matrix of the ith layer the differentiated matrix

$$\frac{\partial \mathbf{M}_i}{\partial \mathrm{QWOT}_i} = \begin{bmatrix} \dfrac{\partial(\cos \phi_i)}{\partial \mathrm{QWOT}_i} & i \dfrac{\partial[(\sin \phi_i)/n_i]}{\partial \mathrm{QWOT}_i} \\[2ex] i \dfrac{\partial(n_i \sin \phi_i)}{\partial \mathrm{QWOT}_i} & \dfrac{\partial(\cos \phi_i)}{\partial \mathrm{QWOT}_i} \end{bmatrix}$$

$$= \frac{\pi}{2\lambda} \begin{bmatrix} -\sin \phi_i & \dfrac{i \cos \phi}{n_i} \\[2ex] i n_i \cos \phi_i & -\sin \phi_i \end{bmatrix} \quad (11.15)$$

and then multiplying the matrices the same way one would compute the matrix of the multilayer.

11.4. Examples

After having computed the matrix of the partial derivatives, standard computer routines can be used to perform the matrix multiplications, matrix inversion, and matrix-vector multiplications.

In order to stay within the range of linearity of Eq. 11.2 only small steps in ΔT are allowed. Consequently, Eq. 11.10 or 11.13 has to be used repeatedly to effect significant changes.

The choice between the two methods is not strictly a matter of comparing the number of layers with the number of change points. Neither Eq. 11.10 nor Eq. 11.13 requires that all the layers of a design be opened up for alterations. So, if the total number of change points exceeds the number of layers one can make m smaller than k by holding some of the thicknesses constant.

Figure 11.1 shows the starting design and the refined design which we already presented in Fig. 9.12 and Table 9.1. Vertical refining was used.

In order to maintain symmetries it is sometimes better to refine the

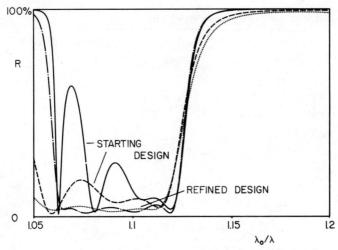

Figure 11.1. Reflectances R_s (solid curve) and R_p (dashed curve) of the design $1 \mid H\ 0.8L\ (0.8L\ HLHLHLH\ 0.8L)^4\ 0.8L\ HLH \mid 1.52$ with $n_H = 2.28$, $n_L = 1.45$, and $\alpha_0 = \beta = 45°$ compared with the reflectances R_s (dash-dotted curve) and R_p (dotted curve) of the first design of Table 9.1.

refractive indices instead of the thicknesses. Instead of the matrix of Eq. 11.15 we have to use

$$\frac{\partial \mathbf{M}_i}{\partial n_i} = \begin{bmatrix} \dfrac{\partial(\cos \phi_i)}{\partial n_i} & i\dfrac{\partial(\sin \phi_i/n_i)}{\partial n_i} \\[2mm] i\dfrac{\partial(n_i \sin \phi_i)}{\partial n_i} & \dfrac{\partial(\cos \phi_i)}{\partial n_i} \end{bmatrix} = \begin{bmatrix} 0 & \dfrac{-i \sin \phi_i}{n_i^2} \\[2mm] i \sin \phi_i & 0 \end{bmatrix} \quad (11.16)$$

In Fig. 11.2 we give an example. Fourteen cycles of horizontal index refining were used. Holding requests ($\Delta T = 0$) were placed at $\lambda_0/\lambda = 0.7$, 0.84, 0.88, 0.93, and 1.0. At $\lambda_0/\lambda = 0.73$ and 0.78 requests of $\Delta T = 0.01\%$ were incorporated.

11.5. Basic Limitations

All partial derivatives must approach zero when R approaches zero. The proof is simple. If any of the partial derivatives is not zero one could generate a negative reflectance with a thickness change of the respective layer. As a consequence, refining methods do not work too well when the desired reflectance is close to zero.

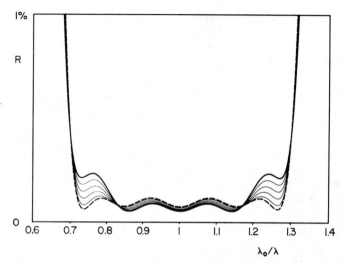

Figure 11.2. Beginning, intermediate, and final reflectance curves of an index-refining process. The design was 1 | L H H1 M1 M2 HH M3 | 1.52. $n_L = 1.45$ and $n_H = 2.4$ were not allowed to change. n_{H1} changed from 2.3 to 2.312, n_{M1} from 1.58 to 1.549, n_{M2} from 1.68 to 1.638, and n_{M3} from 1.78 to 1.772. The solid curve gives the reflectance of the initial design and the dashed curve the reflectance of the final design.

11.6. Problems and Solutions

Problem 11.1

Write a vertical refining program for multilayers at normal and non-normal light incidence. Change requests in both planes of polarization should be worked on simultaneously.

Solution.

Program Listing for Problem 11.1

```
100   ! REFINIG PROGRAM - BAUMEISTER, NAME IS "DROSS_MLRS"
101   OPTION BASE 1
102   ! AFTER PRESSING RUN-KEY THE PROGRAM WAITS FOR THE OPERATOR TO
103   ! PRESS A SOFT KEY TO SELECT BETWEEN LOADING DATA, PLOTTING, OR
104   ! REFINING. THE PROGRAM DOES NOT PLOT TRANSMITTANCE UNLESS THE
105   ! KEY "STARTING DESIGN" WAS PRESSED AND IT DOES NOT REFINE
106   ! UNLESS THE KEY "CORR SPECS" WAS PRESSED
107 Dsgndata:  DATA 1,1.52 ! n(medium),n(substrate)?
108            DATA 2.326,-2.28 ! quot,n (-n for layers to be refined)?
109            DATA .612,-1.45
110            DATA 1.224,-2.28
```

Program Listing for Problem 11.1 (*Continued*)

```
111          DATA .612,-1.45
112          DATA 1.224,-2.28
113          DATA .612,-1.45
114          DATA 2.626,-2.28
115          DATA .6,-1.45
116          DATA 1.2,-2.28
117          DATA .6,-1.45
118          DATA 1.2,-2.28
119          DATA .6,-1.45
120          DATA 2.6,2.28
121          DATA .6,1.45
122          DATA 1.2,2.28
123          DATA .6,1.45
124          DATA 1.2,2.28
125          DATA .6,1.45
126          DATA 2.6,2.28
127          DATA .6,1.45
128          DATA 1.2,2.28
129          DATA .6,1.45
130          DATA 1.2,2.28
131          DATA .6,1.45
132          DATA 2.6,2.28
133          DATA .6,-1.45
134          DATA 1.2,-2.28
135          DATA .6,-1.45
136          DATA 1.2,-2.28
137          DATA .6,-1.45
138          DATA 2.626,-2.28
139          DATA .612,-1.45
140          DATA 1.224,-2.28
141          DATA .612,-1.45
142          DATA 1.224,-2.28
143          DATA .612,-1.45
144          DATA 2.326,-2.28
145          DATA 1,-1.45
146          DATA 1,-2.28
147          DATA 0,0
148 Anglesdata:DATA 45,45 ! angle(incidence),angle(match)?
149 Plotsweepdata: DATA .9,.002,1 ! W-start,W-incr,W-final?
150                DATA 80,100 ! Tmin,Tmax?
151 ! at normal incidence, divide change-points between s- and p-plane
152 Corrspecsdata:  DATA .922,.1 ! change-points,change-amount in s-plane?
153                 DATA .929,.1
154                 DATA .942,.1
```

Program Listing for Problem 11.1 (*Continued*)

```
155                DATA .955,.1
156                DATA 1.005,.2
157                DATA 1.014,.2
158                DATA 0,0
159                DATA .926,.1 ! p-plane
160                DATA .937,.1
161                DATA .955,.2
162                DATA .973,.2
163                DATA .996,.2
164                DATA 1.008,.2
165                DATA 0,0
166    DIM Deriv(30,50),Sqmat(30,30),Corrwn(30)
167    DIM Corram(30),Rsltch(30),Qwot(100),Index(100)
168    DIM Imod(100),Qmod(100),Sqmod(100),Qfactor(100)
169    DEG
170 Start:    DISP "SELECT SOFT KEY"
171           ON KEY 1 LABEL "STARTING DESIGN" GOTO Designin
172           ON KEY 2 LABEL "CORR SPECS" GOTO Corrspecsin
173           ON KEY 3 LABEL "TRANS PLOTS" GOTO Transplts
174           ON KEY 4 LABEL "DUMP CURVES" GOTO Dumpgraphics
175           ON KEY 5 LABEL "ADD PLOT" GOTO Addplot
176           ON KEY 6 LABEL "CORRECT 3X" GOTO Startcorr
177           ON KEY 7 LABEL "RESET DESIGN" GOTO Resetdsgn
178           ON KEY 8 GOTO Start
179           GOTO Start
180 Dumpgraphics:    ALPHA OFF
181                  DUMP GRAPHICS #9
182                  GOTO Start
183 Designin: ALPHA OFF
184           RESTORE Dsgndata
185           READ Nmed,Nsub
186           Ccount=0
187           Lcount=0
188           FOR I=1 TO 100
189           READ Qwot(I),Index(I)
190           IF Qwot(I)=0 THEN Anglesin
191           Lcount=Lcount+1
192           IF Index(I)<0 THEN Ccount=Ccount+1
193           NEXT I
194           DISP "TOO MANY LAYERS"
195 Anglesin: IF Ccount>50 THEN 197
196           GOTO 198
197           DISP "TOO MANY LAYERS TO BE REFINED"
198           RESTORE Anglesdata
```

Program Listing for Problem 11.1 (*Continued*)

```
199          READ Ainc,Amatch
200 Printdsgn: PRINTER IS 9
201          PRINT CHR$(27)&"E"
202          PRINT "     nM=";Nmed
203          PRINT "     nS=";Nsub
204          PRINT
205          PRINT "       qwot      n"
206          FOR I=1 TO 100
207          PRINT USING "5X,DDD.DDD,2X,DDD.DDD";Qwot(I),Index(I)
208          IF Qwot(I)=0 THEN Printang
209          NEXT I
210 Printang: PRINT
211          PRINT "     incid-A=";Ainc
212          PRINT "     match-A=";Amatch
213          PRINT USING "@,#"
214 Modifiers: Nsubm=SQR(1-(Nmed*SIN(Ainc)/Nsub)^2)
215          Nmedm=COS(Ainc)
216          FOR I=1 TO 100
217          IF Qwot(I)=0 THEN Start
218          Imod(I)=SQR(1-(Nmed*SIN(Ainc)/Index(I))^2)
219          Qfactor(I)=Imod(I)/SQR(1-(Nmed*SIN(Amatch)/Index(I))^2)
220          Qmod(I)=Qwot(I)*Qfactor(I)
221          NEXT I
222 Corrspecsin:  ALPHA OFF
223          RESTORE Corrspecsdata
224          Test=0
225          FOR I=1 TO 30
226          READ Corrwn(I),Corram(I)
227          IF Corrwn(I)=0 THEN Test=Test+1
228          IF Test=2 THEN Corrprint
229          NEXT I
230          DISP "TO MANY CORRECTION POINTS"
231 Corrprint: PRINT "    s: corr-wn    corr-am"
232          I=0
233          I=I+1
234          IF Corrwn(I)=0 THEN Pplane
235          PRINT USING "5X,DDD.DDD,2X,DDD.DDD";Corrwn(I),Corram(I)
236          GOTO 233
237 Pplane:  PRINT "    p: corr-wn    corr-am"
238          I=I+1
239          IF Corrwn(I)=0 THEN Rdim
240          PRINT USING "5X,DDD.DDD,2X,DDD.DDD";Corrwn(I),Corram(I)
241          GOTO 238
242 Rdim:  PRINT USING "@,#"
```

Program Listing for Problem 11.1 (Continued)

```
243        Dcount=I-2
244        REDIM Deriv(Dcount,Ccount)
245        REDIM Sqmat(Dcount,Dcount)
246        GOTO Start
247 Transplts:  RESTORE Plotsweepdata
248             READ Wstart,Wincr,Wfinal
249             READ Tmin,Tmax
250             GINIT
251             GRAPHICS ON
252             CSIZE 5,.8
253             ALPHA OFF
254             MOVE 10,87
255             LABEL USING "DDD";Tmax
256             MOVE 12,20
257             LABEL USING "DD";Tmin
258             MOVE 15,65
259             LABEL "T"
260             MOVE 118,15
261             LABEL USING "DD.DD";Wfinal
262             MOVE 15,15
263             LABEL USING "DD.DD";Wstart
264             MOVE 60,15
265             LABEL "WAVENUMBER"
266             VIEWPORT 23,123,20,90
267             WINDOW Wstart,Wfinal,Tmin,Tmax
268             A=(Wfinal-Wstart)/10
269             B=(Tmax-Tmin)/5
270             GRID A,B,Wstart,Tmin
271             GOTO Addplot
272 Resetdsgn:  FOR I=1 TO 100
273             Qmod(I)=Sqmod(I)
274             NEXT I
275             GOTO Pplane
276 Startcorr:  ALPHA OFF
277             Corrcycle=1
278             FOR I=1 TO 100
279             Sqmod(I)=Qmod(I)
280             NEXT I
281             Calcmode=2
282             GOTO 285
283 Addplot:   ALPHA OFF
284             Calcmode=1
285             Polmode=1
286             Corrmode=1
```

Program Listing for Problem 11.1 (*Continued*)

```
287 Pentry:    Dotcount=1
288            ON Calcmode GOTO 289,291      ! =1 Plots, =2 Correction
289            Wrun=Wstart
290            GOTO 295
291            Ip=1  ! Ip is counter for wavenumber position of correction
292            Ipp=1 ! same as Ip but skipps Corrwn=0 positions
293            PRINT USING "/"
294            PRINT "      W-point    T-change"
295            ON Polmode GOTO 296,299
296            Nmedr=Nmed*Nmedm
297            Nsubr=Nsub*Nsubm
298            GOTO 301
299            Nmedr=Nmed/Nmedm
300            Nsubr=Nsub/Nsubm
301            ON Calcmode GOTO 306,302
302            Wrun=Corrwn(Ip)
303            Corrmode=1 ! =1 calculate transmittance at Corr-points
304                       ! =2 calculate derivatives, look for diff-layer
305                       ! =3 finish up derivatives, ignore diff-layers
306 Nextwn:    M11=1
307            M12=0
308            M21=0
309            M22=1
310            Il=1        ! Il: Index for every layer in design
311            Ic=1        ! Ic: Index for layer to be differentiated
312 Loop:      ON Corrmode GOTO 328,319,328
313 Ret:       M11=Retm11
314            M12=Retm12
315            M21=Retm21
316            M22=Retm22
317            Il=Retcount
318            GOTO 328
319            IF Index(Il)>0 THEN 328
320            Retcount=Il
321            Corrmode=3
322            Retm11=M11
323            Retm12=M12
324            Retm21=M21
325            Retm22=M22
326            Phthickn=90
327            GOTO 329
328            Phthickn=0
329            Phthickn=Phthickn+Wrun*Qmod(Il)*90
330            ON Polmode GOTO 331,333
```

Program Listing for Problem 11.1 (Continued)

```
331            Indexr=ABS(Index(I1)*Imod(I1))
332            GOTO 334
333            Indexr=ABS(Index(I1)/Imod(I1))
334            L11=COS(Phthickn)
335            L21=SIN(Phthickn)
336            L12=L21/Indexr
337            L21=L21*Indexr
338            A=M11*L11-M12*L21
339            M12=M11*L12+M12*L11
340            M11=A
341            B=M21*L11+M22*L21
342            M22=M22*L11-M21*L12
343            M21=B
344            I1=I1+1
345            IF I1<=Lcount THEN Loop
346            ON Calcmode GOTO Trans,347
347            ON Corrmode GOTO 348,Nextd,Deriv
348            T11=M11
349            T12=M12
350            T21=M21
351            T22=M22
352 Trans:    A=Nmedr*M11/Nsubr+M22
353            B=Nmedr*M12+M21/Nsubr
354            Trans=A*A+B*B
355            Trans=400*Nmedr/Nsubr/Trans
356            ON Calcmode GOTO 357,369
357            PLOT Wrun,Trans,1
358            Dotcount=Dotcount+1
359            ON Polmode GOTO 364,360
360            IF Dotcount>3 THEN
361            PENUP
362            Dotcount=1
363            END IF
364            Wrun=Wrun+Wincr
365            IF Wrun<=Wfinal THEN Nextwn
366            PENUP
367            Polmode=Polmode+1
368            ON Polmode GOTO Pentry,Pentry,Start
369            PRINT USING "5X,DDD.DDD";Wrun,Trans
370            Corrmode=2
371            GOTO Nextwn
372 Deriv:    A=Nmedr*M11*T11/Nsubr+Nsubr*M22*T22/Nmedr
373            B=Nsubr*Nmedr*M12*T12+M21*T21/Nmedr/Nsubr
374            Deriv(Ipp,Ic)=-(A+B)*Trans*Trans*PI*Wrun*Qmod(Retcount)/40000
```

Program Listing for Problem 11.1 (*Continued*)

```
375         Ic=Ic+1
376         Corrmode=2
377         GOTO Ret
378 Nextd:  Ip=Ip+1
379         Ipp=Ipp+1
380         IF Corrwn(Ip)=0 THEN
381         Polmode=Polmode+1
382         Ip=Ip+1
383         IF Polmode>2 THEN Matrixinv
384         END IF
385         GOTO 295
386 Matrixinv:  I1=1
387             IF I1>Dcount THEN 402
388             I2=1
389             IF I2>I1 THEN 400
390             Sum=0
391             I3=1
392             IF I3>Ccount THEN 396
393             Sum=Sum+Deriv(I1,I3)*Deriv(I2,I3)
394             I3=I3+1
395             GOTO 392
396             Sqmat(I1,I2)=Sum
397             Sqmat(I2,I1)=Sum
398             I2=I2+1
399             GOTO 389
400             I1=I1+1
401             GOTO 387
402             MAT Sqmat= INV(Sqmat)
403             PRINT USING "/"
404             PRINT "         n          QWOT"
405             I3=1
406             IF I3>Ccount THEN 425
407             I1=1
408             IF I1>Dcount THEN 412
409             Rsltch(I1)=Deriv(I1,I3)
410             I1=I1+1
411             GOTO 408
412             I2=1
413             IF I2>Dcount THEN 423
414             Sum=0
415             I1=1
416             IF I1>Dcount THEN 420
417             Sum=Sum+Rsltch(I1)*Sqmat(I1,I2)
418             I1=I1+1
```

Program Listing for Problem 11.1 (Continued)

```
419          GOTO 416
420          Deriv(I2,I3)=Sum
421          I2=I2+1
422          GOTO 413
423          I3=I3+1
424          GOTO 406
425          I3=1
426          IF I3>Ccount THEN 443
427          Sum=0
428          Polmode=1
429          I1=1
430          I4=1
431          IF Corrwn(I4)=0 THEN
432              Polmode=Polmode+1
433              IF Polmode>2 THEN 440
434                  I4=I4+1
435          END IF
436          Sum=Sum+Deriv(I1,I3)*Corram(I4)
437          I1=I1+1
438          I4=I4+1
439          GOTO 431
440          Rsltch(I3)=Sum
441          I3=I3+1
442          GOTO 426
443          I1=1
444          I2=1
445          IF I1>Lcount THEN 452
446          IF Index(I1)>0 THEN 450
447          Qmod(I1)=Qmod(I1)*(1+Rsltch(I2)/100)
448          PRINT USING "5X,DDD.DDDD";-Index(I1),Qmod(I1)/Qfactor(I1)
449          I2=I2+1
450          I1=I1+1
451          GOTO 445
452          Corrcycle=Corrcycle+1
453          IF Corrcycle<4 THEN 285
454          GOTO Start
455          END
```

Problem 11.2

For the refinement of the starting design of Fig. 11.2, use index instead of thickness refining. Also, use the same change requests.

Solution. The result is shown in Fig. 11.3. Because of the symmetry associated with equal thickness designs, index refining worked over the full range in Fig. 11.2 from 0.7 to 1.3, inspite of the fact that change requests were only made for $\lambda_0/\lambda < 1$. This benefit is lost with thickness refining.

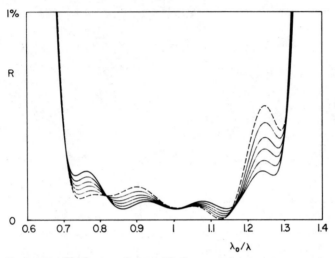

Figure 11.3. Thickness refining with the same change requests and for the same design as in Fig. 11.2. Starting design is $1 \mid \text{L H H1 M1 M2 HH M3} \mid 1.52$ with $n_\text{L} = 1.45$, $n_\text{H} = 2.4$, $n_{\text{H1}} = 2.3$, $n_{\text{M1}} = 1.58$, $n_{\text{M2}} = 1.68$, and $n_{\text{M3}} = 1.78$. Final design was $1 \mid \text{L H 1.014H1 0.994M1 0.979M2 H 1.029M3} \mid 1.52$ (dashed curve).

Producibility of Designs

Often, multilayer designs look good on paper but are almost impossible to produce. Aside from too high a number of layers, this is the case when

1. The design is intended for spectral regions where the dispersion of the refractive indices of one or all the used coating materials is high.
2. All or some of the layers are *inhomogeneous* (the refractive index varies as a function of thickness within one layer).
3. All or some of the layers absorb or scatter excessively.
4. The design is too sensitive to variations of the thickness and/or refractive indices of the individual layers.

As mentioned in Chap. 1, the in-depth treatment of these topics is beyond the scope of this book and only sample solutions can be given.

12.1. Dispersion

The dispersion of the refractive indices is seldom significant enough to be an important consideration in the initial design process. But it can in some cases be very annoying and might require redesign, change of materials, or special manufacturing procedures (see Fig. 12.2).

For antireflection coatings, high dispersion means a decrease in bandwidth. It acts like a progressive increase in optical thickness on the short wavelength side. In Fig. 12.1 we show the reflectance of a design which is similar to one of the designs given in Fig. 4.10, but this time with dispersion. The reduction in bandwidth is 5 percent. For edge filters, dispersion can generate the so-called half-wave hole which we already encountered in Fig. 9.2. In the presence of dispersion a quarter-wave film at λ_0 is no longer a half-wave film at $\lambda_0/2$. As a consequence the second-order reflectance band is no longer suppressed (Chap. 8). The half-wave hole can be avoided by making sure that the two materials of the alternating stack are one-half-wave thick at the same position. The fact that they now are no longer equally thick at the quarter-wave position is of little consequence. Figure 12.2 gives an example.

12.2. Inhomogeneity

Inhomogeneity is mostly a problem in the production of multilayer antireflection coatings. Recently, DeBell[1] investigated its influence on the half-wave layer of a quarter/half/quarter design (Fig. 4.7). By replacing the half-wave layer with four eighth-wave layers having increasing refractive indices (towards the substrate) he found that the

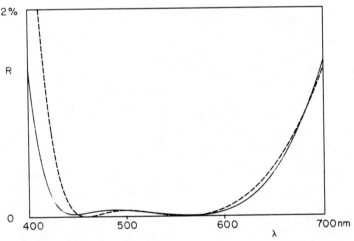

Figure 12.1. Simulation of the effect of dispersion in an antireflection coating design. The index sequence is 1 (medium), 1.45 (SiO$_2$), 2.48 (solid curve) or 2.63/350 nm, 2.58/400 nm, 2.53/450 nm, 2.48/500 nm, 2.43/550 nm, 2.39/600 nm, 2.36/650 nm, 2.3/1200 nm (dashed curve), 1.75, 1.52 (glass). The physical thickness sequence is 87.9 nm, 102.8 nm, and 75 nm.

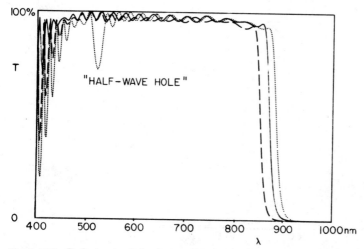

"HALF – WAVE HOLE"

Figure 12.2. Reflectance of the design of Fig. 6.12 ($p = 3$; $n_L = 1.45$). For the high-index material we assumed $n_H = 2.32$ (solid curve) or the same dispersion as in Fig. 12.1. For the optical thicknesses we assumed a match at 500 nm (dotted curve) or 1000 nm (dashed curve) ($\lambda_0 = 1000$ nm).

bandwidth was widened and the reflectance in the center substantially increased. By increasing the refractive index of the layer next to the substrate from 1.63 to 1.7, he could eliminate the reflectance increase in the center. The resulting coating had as low a reflectance as the homogeneous quarter/half/quarter coating but had a wider low-reflectance zone. This was exactly the case with the experimentally produced coatings by Cox, Hass, and Thelen.[2] The higher than intended refractive index of CeF_3 can easily be attributed to CeO_2 contamination (Fig. 12.3).

Another way of dealing with this inhomogeneity was reported by Apfel and Snavely.[3] They broke up the half wave into two slightly thinner quarter waves and inserted a thin layer of MgF_2.

12.3. Absorption and Scatter

Residual absorption and scatter are a major problem with laser coatings and especially high reflecting stacks (Fig. 5.2). Let us now assume that some of the layers have nonzero absorption ($n \rightarrow n - ik$ with $k \neq 0$). The reflectance and the overall absorptance of the stack depend very much on the location and the relative thickness of the absorbing layers. The further away from the light incidence side they are (Koppelmann[4]) and the smaller the relative thickness is (Lissberger[5]), the higher is the reflectance and the lower is the absorptance. In Fig. 12.4 we assume four of the high-index layers to be absorbing. Relo-

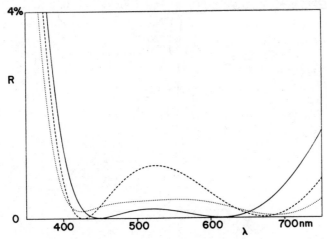

Figure 12.3. Effect of inhomogeneity in the middle layer of a $\lambda/4$–$\lambda/2$–$\lambda/4$ antireflection coating: $1.0 \mid L\ H1/2\ H2/2\ H3/2\ H4/2\ M \mid 1.52$ with $n_L = 1.38$, $n_{H1} = 1.96$, $n_{H2} = 2.02$, $n_{H3} = 2.08$, $n_{H4} = 2.135$, $n_{M1} = 1.63$ (dashed curve), and $n_{M2} = 1.7$ (dotted curve) compared to $1.0 \mid L\ H2\ H2\ M1 \mid 1.52$ (solid curve) ($\lambda_0 = 520$ nm).

cating the absorbing layers from the side next to the light entry to the opposite side reduces $(100\% - R)$ to one-half. The decrease in absorptance $(100\% - R - T)$ is higher.

Figure 12.5 gives the dependence on relative thickness. Starting with a stack of layers with equal optical thickness, the reflectance is in-

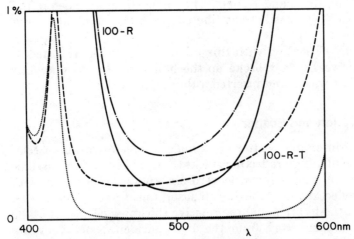

Figure 12.4. Reflectance $(100\% - R)$ and transmittance $(100\% - R - T)$ of the designs $1 \mid (H1\ L)^4\ (H2\ L)^3\ H2 \mid 1.52$ (dashed and dash-dotted curves) and $1 \mid (H2\ L)^4\ (H1\ L)^3\ H1 \mid 1.52$ (solid and dotted curves) with $n_{H1} = 2.35 - 0.001i$, $n_{H2} = 2.35$, and $n_L = 1.45$ ($\lambda_0 = 500$ nm).

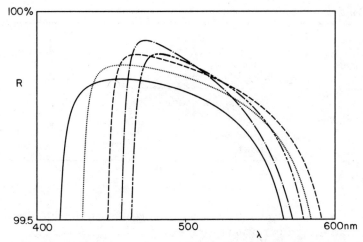

Figure 12.5. Reflectance of the designs $1 \mid (a\text{H } b\text{L})^{13} a\text{H} \mid 1.51$ with $n_\text{H} = 2.35 - 0.001i$ and $n_\text{L} = 1.35$ ($\lambda_0 = 500$ nm): $b/a = 1$ (solid curve), $b/a = 1.36$ (dotted curve, $b/a = 1.87$ (dashed curve), $b/a = 2.78$ (dash-dot-dotted curve), and a and b continuously variable (from Carniglia and Apfel[6]) (dash-dotted curve).

creased if we decrease the relative thickness of the absorbing layer (Lissberger[5]). Carniglia and Apfel[6] show that the reflectance can be further increased by using continuously variable *optimum pairs* of layers.

12.4.   Variations in Thickness and/or Refractive Index

The most important criterion for the producibility of a design is its sensitivity to its main optical parameters, the thicknesses and/or refractive indices of its individual layers. Thoeni[7] has pointed out that some design types are more sensitive to thickness variations while others are more sensitive to index variations. Table 12.1 gives a survey.

Refractive index variations are normally the effect of increases (decreases) of temperature, pressure, or gas and do not have the statistical nature of optical thickness variations. Their effect is best analyzed by determining their most probable values and calculating the resulting reflectance changes (Sec. 12.2).

This is different for optical thickness variations. Their variations are mostly statistical and their influence on the reflectance can be determined with Monte Carlo calculations.

In the following calculations we used a random generator to vary the optical thicknesses within the maximum given error limits while

TABLE 12.1   Sensitivity to Optical Parameters of Some Coating Types

More sensitive to refractive index variations	Equally sensitive to both refractive index and optical thickness variations	More sensitive to optical thickness variations
Polarizers	Multilayer AR coatings	Narrow-band filters
Neutral beam splitters	Edge filters	
Conversion filters	Broad-band coatings	
	Laser coatings	

From Thoeni.[7]

retaining the maximum and minimum reflectance values. One hundred to one thousand design variations were evaluated.

In Fig. 12.6 we show the effect of a ±3 percent error on a standard four-layer antireflection coating. It appears to be an acceptable production tolerance.

For the edge filter of Fig. 12.7 a ±1 percent error produced a ±0.4 percent edge variation and a reflectance increase of about 6 percent.

Unexpected relationships are revealed in comparing the sensitivity to thickness variations of two narrow bandpass designs (Fig. 10.22). The second design was derived out of the first design by adding a half-

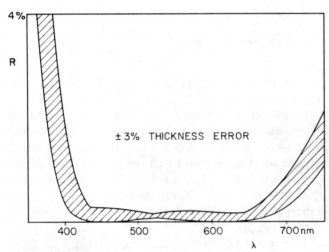

Figure 12.6. Effect of optical thickness variations on the two-material four-layer antireflection coating $1 \mid L\ 2.108H\ 0.331L\ 0.225H \mid 1.52$ with $n_L = 1.38$ and $n_H = 2.08$ ($\lambda_0 = 520$ nm). The optical thicknesses were varied with a random generator between ±3 percent limits. A set of 100 cases was evaluated.

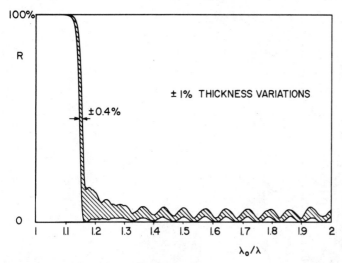

Figure 12.7. Effect of optical thickness variations on the edge filter
1 | 1.12(L/2 H L/2) 1.06(L/2 H L/2) 1.03 (L/2 H L/2) 1.015(L/2 H L/2) (L/
2 H L/2)^5 1.015 (L/2 H L/2) 1.03(L/2 H L/2) 1.06(L/2 H L/2)   1.12(L/
2 H L/2) | 1.52 with n_L = 1.45 and n_H = 2.35. Thicknesses were var-
ied within ± 1 percent with a random generator. A set of 1000 cases
was evaluated.

wave-thick film in the center. The filter became slightly narrower and
the two 3 percent dips in the passband were eliminated. But, if we now
try to produce the design we will have a worse filter if our error tol-
erances are ± 1 percent, an equally good filter for ± 0.5 percent, and
a better filter with ± 0.25 percent (Fig. 12.8).

For the interpretation of Figs. 12.6, 12.7, 12.8, and 12.10 it is necessary to realize that the shaded areas do not contain all possible
reflectance values—they contain only those values which we can expect
with a yield of 99.9 percent.

12.5. Refractive Indices of Available Coating Materials

In general, optical constants are determined with very high precision,
e.g., the refractive indices of optical glasses are listed to five decimal
places. This is quite different in thin film optics. Because the film
structure is heavily dependent on the deposition conditions (rate of
deposition, residual gas, starting material composition, degree of ion-
ization, etc.), it is customary to give the refractive indices of coating
materials only to two decimal places.

When thermal evaporation out of incandescent sources (such as tung-
sten boats) was the only means to deposit thin films, only a handful

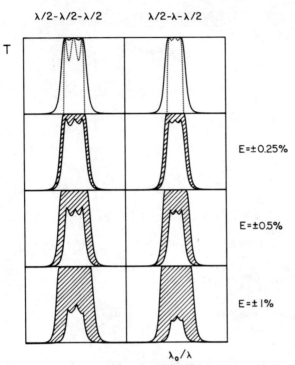

λ/2-λ/2-λ/2   λ/2-λ-λ/2

T

E=±0.25%

E=±0.5%

E=±1%

λ₀/λ

Figure 12.8. Comparison of the sensitivity-to-thickness vari-
ations of two narrow bandpass designs: 1.45 | HLHLH LL
HLHLHLHLHLH LL HLHLHLHLHLH LL HLHLH | 1.45
(left side) and 1.45 | HLHLH LL HLHLHLHLHLH LLLL
HLHLHLHLHLH LL HLHLH | 1.45 (right side) with n_L =
1.45 and n_H = 2.35. Transmittances are plotted in the top
row [0–100 percent (solid curve), 90–100 percent (dotted
curve)]. In the other rows are plotted possible maximum and
minimum transmittances (0–100 percent) for random thickness variations of ±1 percent (bottom row), ±0.5 percent
(second row from bottom), and ±0.25 percent (second row
from top) (0.96 ≤ λ_0/λ ≤ 1.04). A set of 1000 cases was evaluated.

of materials was available which would deposit into clear, stable, and
reasonbly hard films. Electron beam evaporation has increased this
number considerably. Sputtering, plasma-enhanced chemical vapor
deposition, and other emerging technologies are expanding it further.

In Table 12.2 are listed the ranges of transparency and the approximate refractive indices of the "classical" coating materials. Around
80 percent of all optical coatings currently in existence use these materials.

TABLE 12.2   Ranges of Transparency and Refractive Indices of Some Classical Coating Materials

Material	Refractive index at the λ below (in μm)					Range of transparency (in μm)
	0.35	0.55	1.0	1–8	10	
Ge				4.2	4.2	1.7–25
Si			3.9	3.42		1.0–9
TiO_2		2.32	2.20			0.4–6
ZnS		2.36	2.27	2.24		0.4–14
ZrO_2	2.15	2.05	2.0			0.3–
Ta_2O_3	2.31	2.16	2.09			0.3–
SiO		2.0	1.9	1.85		0.5–8
Al_2O_3	1.66	1.63	1.60			0.15–9
Si_2O_3		1.57	1.55			0.3–8
SiO_2	1.48	1.46	1.45			0.16–8
MgF_2	1.39	1.38		1.36		0.13–10
Cryolite		1.30–1.35				0.13–9

SOURCE: From P. W. Baumeister (1987). Optical Coating Technology. Lecture notes for a five-day short course, Engineering 823.17. (University of California at Los Angeles, University Extension).

12.6   Problems and Solutions

Problem 12.1

Find an easy way of determining whether a coating material deposits inhomogeneously and whether the index increases or decreases with increasing thickness. Assume no absorption.

Solution.   At the half-wave position, a single-layer film should have the same reflectance as the uncoated substrate surface (Problem 2.3). Let us approximate the half-wave film by two quarter-wave films with slightly different refractive indices. It is then obvious that the reflectance is lower when the index next to the substrate is higher and higher when the index is lower.

Problem 12.2

Let us assume we have three coating materials available to design a 15-layer-high reflector. The refractive indices are $n_{H1} = 2.35 - 0.001i$, $n_{H2} = 2.05$, and $n_L = 1.45$. Which combination of the three materials gives the highest reflectance?

Solution.   In Fig. 12.9 the first, second, and third H1-layer is replaced by an H2-layer. Replacing the first H1-layer brings, inspite of the lower

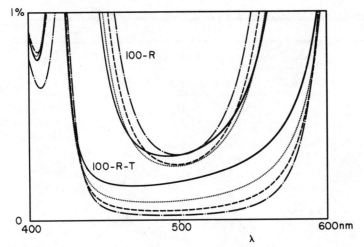

Figure 12.9. Reflectance $(100\% - R)$ and absorptance $(100\% - R - T)$ of the designs $1 \mid (H1\ L)^7\ H1 \mid 1.52$ (solid curve), $1 \mid H2\ (L\ H1)^7 \mid 1.52$ (dotted curve), $1 \mid H2\ L\ H2\ (L\ H)^6 \mid 1.52$ (dashed curve), and $1 \mid H2\ L\ H2\ L\ H2\ (L\ H1)^5 \mid 1.52$ (dash-dotted curve) with $n_{H1} = 2.35 - 0.001i$, $n_{H2} = 2.05$, and $n_L = 1.45$ ($\lambda_0 = 500$ nm).

refractive index, a rather significant increase in reflectance while reducing the absorptance to one-half. Replacing the second H1-layer does not increase the reflectance further but reduces the absorptance again to one-half. Further replacements have a negative influence on the reflectance.

Problem 12.3

Compare the sensitivity to thickness variation of a two-half-wave-spacer narrow bandpass filter with that of a three-spacer filter.

Solution. From Fig. 12.10 we can tell that the higher performance of the three-spacer filter requires a much better thickness control. For a peak transmittance of 80 percent the two-half-wave filter gets by with a 1 percent thickness tolerance while the three-half-wave filter requires 0.5 percent.

Problem 12.4

In Fig. 4.24 we showed two dual-band antireflection coatings by Szafranek and Lubezky.[8] Only the minimum and maximum refractive indices could be selected freely—the other two followed from the theory.

2 CAVITIES   3 CAVITIES

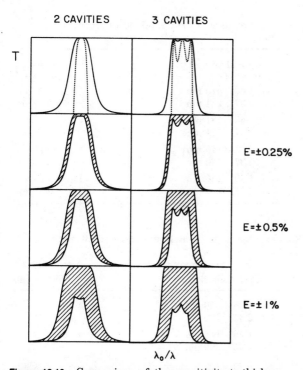

T

E=±0.25%

E=±0.5%

E=±1%

λ_0/λ

Figure 12.10. Comparison of the sensitivity-to-thickness variations of two narrow bandpass designs: 1.45 | HLHLH LL HLHLHLHLHLH LL HLHLHLHLHLH LL HLHLH | 1.45 (right side) and 1.45 | HLHLH LL HLHLHLHLHLH LL HLHLH | 1.45 (left side) with n_L = 1.45 and n_H = 2.35. Transmittances are plotted in the top row [0–100 percent (solid curve), 90–100 percent (dotted curve)]. In the other rows are plotted possible maximum and minimum transmittances (0–100 percent) for random thickness variations of ±1 percent (bottom row), ±0.5 percent (second row from bottom), and ±0.25 percent (second row from top) ($0.96 \leq \lambda_0/\lambda \leq 1.04$). A set of 1000 cases was evaluated.

Convert the four-material four-layer design into a two-material six-layer design.

Solution. In their paper, Szafranek and Lubetzky[8] offered two versions of each design: one replacing the in-between refractive indices by equivalent layers and the other by further computer refining after incorporating dispersion. Tables 12.3 and 12.4 give the designs.

TABLE 12.3   Two-Material Six-Layer Versions of the Dual-Band Antireflection
Coating of Fig. 4.24 with n = 1.40

Original design	Equivalent design	Refined design
1.40, 570 nm	1.40, 570 nm 2.37, 211 nm	1.40, 609 nm 2.37, 278 nm
1.95, 570 nm	1.40, 131 nm 2.37, 211 nm	1.40, 117 nm
2.37, 1140 nm	2.37, 1140 nm 2.37, 180 nm	2.37, 1447 nm
1.90, 514 nm	1.40, 136 nm 2.37, 180 nm	1.40, 149 nm 2.37, 173 nm

TABLE 12.4   Two-Material Six-Layer Versions of the Dual-Band Antireflection
Coating of Fig. 4.24 with n = 1.46

Original design	Equivalent design	Refined design
1.46, 550 nm	1.46, 550 nm 2.45, 209 nm	1.46, 601 nm 2.45, 294 nm
2.05, 550 nm	1.46, 118 nm 2.45, 209 nm	1.46, 107 nm
2.45, 1110 nm	2.45, 1110 nm 2.45, 165 nm	2.45, 1443 nm
1.93, 490 nm	1.46, 143 nm 2.45, 165 nm	1.46, 162 nm 2.45, 164 nm

References

Chapter 1

1. P. W. Baumeister (1984). Optical Interference Coating Technology, Lecture Notes for a Five-Day Short Course, Engineering 823.17. (University of California at Los Angeles, University Extension).
2. K. H. Guenther and H. K. Pulker (1976). Electron microscopic investigations of cross sections of optical thin films. *Appl. Opt.,* 15:2992–2997.

Chapter 2

1. E. Hecht and A. Zajac (1974). *Optics.* Addison-Wesley Publishing Co., Boston.
2. P. H. Berning (1963). Theory and calculations of optical thin films. In G. Hass, ed.: *Physics of Thin Films,* vol. 1. Academic Press, New York and London, pp. 69–121.
3. Z. Knittl (1976). *Optics of Thin Films.* John Wiley & Sons, New York.
4. E. Kreyszig (1962). *Advanced Engineering Mathematics.* John Wiley and Sons, New York.
5. S. Thelen (1984). Private communication. Zürich.
6. F. Abelès (1950). Recherches sur la propagation des ondes électromagnétique sinusoidales dans les milieux stratifies. Application aux couches minces. *Ann. de Physique,* 5, 596–640, and 706–782.
7. E. Delano and R. Pegis (1969). Methods of synthesis for dielectric multilayer filters. In E. Wolf, ed.: *Progress in Optics,* vol. 7. North-Holland Publishing Co., Amsterdam, pp. 67–137.
8. C. Lanczos (1964). *Applied Analysis.* Prentice-Hall, New York.
9. M. Abramowitz and I. A. Stegun (1964). *Handbook of Mathematical Functions.* National Bureau of Standards, Applied Mathematics Series 55.
10. L. Young (1961). Spin matrix exponentials and transmission matrices. *Quarterly Applied Math.,* 19:25–30.
11. Technical Information, Color Filter Glass (1982). Schott Optical Glass, Inc., 400 York Ave., Duryea, PA 18642.
12. S. MacNeille (1946). Beamsplitter. United States Patent #2,403,731.
13. A. Thelen (1966). Equivalent layers in multilayer filters. *J. Opt. Soc. Am.,* 56:1533–1538.

Chapter 3

1. A. Herpin (1947). Calcul du pouvoir réflecteur d'un système stratifiè quelconque. *Compt. Rend.,* 225:182.
2. L. I. Epstein (1952). The design of optical filters. *J. Opt. Soc. Am.,* 42:806–810.
2a. A. Thelen (1966). Equivalent layers in multilayer filters. *J. Opt. Soc. Am.,* 56:1533–1538.
3. L. I. Epstein (1955). Improvements in heat-reflecting filters. *J. Opt. Soc. Am.,* 45:360–362.

4. L. I. Epstein (1952). *Theory of Multilayer Coatings.* Bausch and Lomb, Inc., Rochester, New York (manuscript of unpublished book).
5. R. Miller (1976). Private communication, Optical Society of America. Topical Meeting on Optical Interference Coatings, Asilomar Conference Grounds, Pacific Grove, Calif., Feb. 24–26.
6. M. C. Ohmer (1978). Design of three-layer equivalent films. *J. Opt. Soc. Am.,* 68: 137–139.
7. J. A. Dobrowolski and S. H. C. Piotrowski (1982). Refractive index as a variable in the numerical design of optical thin film systems. *Appl. Opt.,* 21:1502–1511.
8. L. I. Epstein (1979). Design of optical filters, pt. 2. *Appl. Opt.,* 18:1478–1479.
9. H. Pohlack (1952). *Die Synthese optischer Interferenzschichtsysteme mit vorgegebenen Spektraleigenschaften.* Jenaer Jahrbuch, pp. 181–221.
10. L. Young (1961). Synthesis of multiple antireflection films over a prescribed frequency band. *J. Opt. Soc. Am.,* 51:967–974.
11. J. S. Seeley (1961). Synthesis of interference filters. *Proc. Phys. Soc.,* 78:998–1008.
12. R. E. Collin (1955). Theory and design of wide-band multisection quarter-wave transformers. *Proc. IRE,* 43:179–185.
13. L. Young (1962). Stepped-impedance transformers and filter prototypes. IRE Transactions on Microwave Theory Tech., vol. MTT-10, pp. 339–359.
14. L. Young (1960). The quarter-wave transformer prototype circuit. IRE Transactions on Microwave Theory Tech., vol. MTT-7, pp. 483–489.
15. R. Levy (1965). Tables of element values for the distributed low-pass prototype filter. IEEE Transactions on Microwave Theory and Techniques, vol. MTT-13, pp. 514–536.
16. H. J. Riblet (1957). General synthesis of quarter-wave impedance transformers. IRE Transactions on Microwave Theory Tech., vol. MTT-5, pp. 36–43.
17. P. I. Richards (1948). Resistor-transmission-line circuits. *Proc. I. R. E.,* 36:217–220.
18. P. Baumeister (1982). Use of microwave prototype filters to design multilayer dielectric bandpass filters. *Appl. Opt.,* 21:2965–2967.
19. W. K. Chen (1976). *Theory and Design of Broadband Matching Networks.* Pergamon Press, Oxford, New York.
20. S. D. Smith (1957). Design of multilayer filters by considering two effective interfaces. *J. Opt. Soc.Am.,* 47:43–50.
21. J. Mouchart (1978). Thin film optical coatings, pt. 5, Buffer layer theory. *Appl. Opt.,* 17:72–75.
22. Z. Knittl (1981). Control of polarization effects by internal antireflection. *Appl. Opt.,* 20:105–110.

Chapter 4

1. K. Schuster (1949). Anwendung der Vierpoltheorie auf die Probleme der optischen Reflexionsminderung, Reflexionsverstärkung und der Interferenzfilter. *Annalen der Physik.,* 6. Folge, Band 4, pp. 352–356.
2. Jenaer Glaswerk Schott & Gen (1943). Ueberzug zur Verminderung der Oberflaechenreflexion, Schweizer Patent 223344, January 4 (based on classified German patent by W. Geffcken, July 18, 1940).
3. A. Musset and A. Thelen (1970). Multilayer antireflection coatings. In E. Wolf, ed.: *Progress in Optics,* vol. VIII. North-Holland Publishing Co., Amsterdam, pp. 201–237.
4. A. W. Louderback and M. A. Zook (1972). Method of applying a multilayer antireflection coating to a substrate. United States Patent #3,695,910, October 3.
5. W. Geffcken (1944). Schicht zur Aenderung des Reflexionsvermoegens abwechselnd uebereinanderliegender Teilschichten aus zwei Stoffen von verschiedener Brechzahl. German Patent #742463, January 18.
5a. P. W. Baumeister (1984). Optical Interference Coating Technology, Lecture Notes for a Five-Day Short Course, Engineering 823.17 (University of California at Los Angeles, University Extension).

6. A. Thelen (1974). Reflection-reducing coating. United States Patent #3,854,796, December 17.
7. I. Szafranek and I. Lubezky (1983). Broad, Double-Band Antireflection Coatings on Glasses for 1.06 μm and Visible or Ultraviolet Radiation: Design and Experiment. Proceedings of the SPIE, vol. 401, pp. 138–146.
8. A. Thelen (1960). Theoretical studies on multilayer antireflection coatings (abstract). *JOSA*, 50:509.
9. C. J. Snavely, I. Lewin, and E. A. Small (1979). Lighting fixtures and glass enclosures having high angle anti-reflecting coatings. United States Patent #4,173,778, November 6.
10. J. A. Dobrowolski (1975). Modern Computational Methods for Optical Thin Film Systems. 3d International Conference on Thin Films, Budapest, Hungary, August 23–29.
11. H. A. Macleod (1978). A new approach to the design of metal-dielectric thin-film optical coatings. *Optica Acta*, 25:93–106.
12. A. Thelen (1974). Antireflective multilayer coating on a highly refractive substrate. United States Patent #3,829,197, August 13.

Chapter 5

1. G. Hass and N. W. Scott (1949). Silicon monoxide protected front surface mirrors. *J. Opt. Soc. Am.*, 39:179–184.
2. G. Hass (1955). Filmed surfaces for reflecting optics. *J. Opt. Soc. Am.*, 45:945–952.
3. H. Pohlack (1964). Lichtdurchlässiger Spiegel. German Patent DBP 1013089, April 9.
4. P. W. Baumeister and J. M. Stone (1956). Broad-band multilayer film for Fabry-Perot interferometers. *JOSA*, 46:228–229.
5. W. Lobsiger (1977). Polarisationseffekte in dielektrischen Dünnschicht-Systemen. Proceedings Laser 77, Munich, pp. 20–24.
6. V. R. Costich (1970). Reduction of polarization effects in interference coatings. *Appl. Opt.*, 9:866–870.
7. Z. Knittl and H. Houserková (1982). Equivalent layers in oblique incidence: The problem of unsplit admittance and depolarization of partial reflectors. *Appl. Opt.*, 21:2055–2068.
8. P. Baumeister (1961). The transmittance and degree of polarization of quarter-wave stacks at non-normal incidence. *Optica Acta*, 8:105–119.
9. A. Thelen (1976). Nonpolarizing interference filters inside a glass cube. *Appl. Opt.*, 15:2983–2985.
10. C. M. de Sterke, C. J. van der Laan, and H. J. Frankena (1983). Nonpolarizing beam splitter design. *Appl. Opt.*, 22:595–601.
11. J. Mouchart (1978). Thin film optical coatings. pt. 5. Buffer layer theory. *Appl. Opt.*, 17:72–75.
12. Z. Knittl (1981). Control of polarization effects by internal antireflection. *Appl. Opt.*, 20:105–110.

Chapter 7

1. A. Thelen (1971). Design of optical minus filters. *J. Opt. Soc. Am.*, 61:365–369.
2. P. Baumeister (1981). Theory of rejection filters with ultranarrow bandwidths. *J. Opt. Soc. Am.*, 71:604–606.

Chapter 8

1. L. I. Epstein (1955). Improvements in heat-reflecting filters. *J. Opt. Soc. Am.*, 45: 360–362.

2. A. Thelen (1963). Multilayer filters with wide transmittance bands. *J. Opt. Soc. Am.*, 53:1266–1270.
3. A. Thelen (1973). Multilayer with wide transmittance bands, pt. II. *J. Opt. Soc. Am.*, 63:65–68.

Chapter 9

1. S. MacNaille (1946). Beamsplitter. United States Patent #2,403,731.
1a. M. Banning (1947). Practical methods of making and using multilayer filters. *J. Opt. Soc. Am.*, 37:792–797.
2. A. Thelen (1981). Nonpolarizing edge filters. *J. Opt. Soc. Am.*, 71:309–314.
2a. A. Thelen (1966). Equivalent layers in multilayer filters. *J. Opt. Soc. Am.*, 56:1533–1538.
3. A. Thelen (1984). Nonpolarizing edge filters, pt. 2. *Appl. Opt.*, 23:3541–3543.

Chapter 10

1. W. Geffcken (1939). Interferenzlichtfilter, DRP 716 153, Schott & Gen., patented in Germany starting December 8.
2. C. Fabry and A. Perot (1899). Théorie et applications d'une nouvelle méthode de spéctroscopie interférentielle. *Ann. Chim. Phys.*, 7:115–144.
3. W. Geffcken (1944). Interferenzlichtfilter, DBP 91 30 05, Schott & Gen., patented in West Germany starting November 15.
4. H. D. Polster (1952). A symmetrical all-dielectric interference filter. *J. Opt. Soc. Am.*, 42:21–24.
5. A. F. Turner (1952). Wide band pass multilayer filters. *J. Opt. Soc. Am.*, 42:878(A).
5a. S. D. Smith (1957). Design of multilayer filters by considering two effective interfaces. *J. Opt. Soc. Am.*, 47:43–50.
5b. A. Thelen (1966). Equivalent layers in multilayer filters. *J. Opt. Soc. Am.*, 56:1533–1538.
6. C. Jacobs (1981). Dielectric square bandpass design. *Appl. Opt.*, 20:1039–1042.
6a. R. Levy (1965). Tables of element values for the distributed low-pass prototype filter. IEEE Transactions on Microwave Theory and Techniques, vol. MTT-13, pp. 514–536.
6b. P. Baumeister (1982). Use of microwave prototype filters to design multilayer dielectric bandpass filters. *Appl. Opt.*, 21:2965–2967.
7. P. Baumeister (1983). Simplified equations for maximally flat all-dielectric bandpass design. *Appl. Opt.*, 22:1960.

Chapter 11

1. P. W. Baumeister and J. M. Stone (1956). Broad-band multilayer film for Fabry-Perot interferometers. *J. Opt. Soc. Am.*, 46:228–229.
2. J. A. Dobrowolski (1965). Completely automatic synthesis of optical thin film systems. *Appl. Opt.*, 4:937–946.
3. A. L. Bloom (1981). Refining and optimization in multilayers. *Appl. Opt.*, 20:66–73.
4. H. M. Lidell (1981). *Computer-Aided Techniques for the Design of Multilayer Filters*. Adam Hilger Ltd., Bristol.
5. S. Rosen and C. Eldert (1954). Least-squares method for optical correction. *J. Opt. Soc. Am.*, 44:250–252.
6. P. Baumeister (1958). Design of multilayer filters by successive approximations. *J. Opt. Soc. Am.*, 48:955–958.

Chapter 12

1. G. W. DeBell (1983). Antireflexion Coatings Utilizing Multiple Half Waves. Proceedings of the SPIE, vol. 401, pp. 127–137.
2. J. T. Cox, G. Hass, and A. Thelen (1962). Triple-layer antireflection coatings on glass for the visible and near infrared. *J. Opt. Soc. Am.*, 52:965–969.
3. J. A. Apfel and C. J. Snavely (1973). Wide band anti-reflection coating and article coated therewith. United States Patent #3,761,160, September 25.
4. G. Koppelmann (1960). On the theory of multilayers consisting of weakly absorbing materials and their use in interferometric mirrors (in German). *Ann. Phys. (Leipzig)*, 5:388–396.
5. P. H. Lissberger (1978). The ultimate reflectance of multilayer dielectric mirrors. *Optica Acta*, 25:291–298.
6. C. K. Carniglia and J. H. Apfel (1980). Maximum reflectance of multilayer dielectric mirrors in the presence of slight absorption. *JOSA*, 70:523–534.
7. W. P. Thoeni (1982). Deposition of optical coatings: Process control and automation. *Thin Solid Films*, 88:385–397.
8. I. Szafranek and I. Lubezky (1983). Broad, Double-Band Antireflection Coatings on Glasses for 1.06 μm and Visible or Ultraviolet Radiation: Design and Experiment. Proceedings of SPIE, vol. 401, pp. 138–146.

INDEX

About the Author

Alfred Thelen, Dr.-Ing., is currently Vice President in charge of Thin Film research at Leybold AG, Hanau, West Germany. Previously, Dr. Thelen held positions with Optical Coating Laboratory, Inc., of Santa Rosa, California, and with Balzers AG in Balzers, Liechtenstein. An internationally respected authority on thin film coatings, he is the author of many journal articles on the subject.